The Algorithmic Unconscious

This book applies the concepts and methods of psychoanalysis to the study of artificial intelligence (AI) and human–AI interaction. It develops a new, more fruitful approach for applying psychoanalysis to AI and machine behavior. It appeals to a broad range of scholars: philosophers working on psychoanalysis, technology, AI ethics, and cognitive sciences, psychoanalysts, psychologists, and computer scientists.

The book is divided into four parts. The first part (Chapter 1) analyzes the concept of "machine behavior." The second part (Chapter 2) develops a reinterpretation of some fundamental Freudian and Lacanian concepts through Bruno Latour's actor-network theory. The third part (Chapters 3 and 4) focuses on the nature and structure of the algorithmic unconscious. The author claims that the unconscious roots of AI lie in a form of projective identification, i.e., an emotional and imaginative exchange between humans and machines. In the fourth part of the book (Chapter 5), the author advances the thesis that neuropsychoanalysis and the affective neurosciences can provide a new paradigm for research on artificial general intelligence.

The Algorithmic Unconscious explores a completely new approach to AI, which can also be defined as a form of "therapy." Analyzing the projective identification processes that take place in groups of professional programmers and designers, as well as the "hidden" features of AI (errors, noise information, biases, etc.), represents an important tool to enable a healthy and positive relationship between humans and AI. Psychoanalysis is used as a critical space for reflection, innovation, and progress.

Luca M. Possati is a researcher in philosophy at the University of Porto, Portugal. His fields of investigation are philosophy of technology and artificial intelligence. He is the author of many studies in phenomenology and the history of philosophy.

ANTINOMIES
Innovations in the Humanities, Social Sciences and Creative Arts

This Series addresses the importance of innovative contemporary, comparative and conceptual research on the cultural and institutional contradictions of our times and our lives in these times. *Antinomies* publishes theoretically innovative work that critically examines the ways in which social, cultural, political and aesthetic change is rendered visible in the global age, and that is attentive to novel contradictions arising from global transformations. Books in the Series are from authors both well-established and early careers researchers. Authors will be recruited from many, diverse countries—but a particular feature of the Series will be its strong focus on research from Asia and Australasia. The Series addresses the diverse signatures of contemporary global contradictions, and as such seeks to promote novel transdisciplinary understandings in the humanities, social sciences and creative arts.

The Series Editor is especially interested in publishing books in the following areas that fit with the broad remit of the series:

- New architectures of subjectivity
- Cultural sociology
- Reinvention of cities and urban transformations
- Digital life and the post-human
- Emerging forms of global creative practice
- Culture and the aesthetic

Series Editor: Anthony Elliott
Hawke Research Institute, University of South Australia

The Algorithmic Unconscious
How Psychoanalysis Helps in Understanding AI
Luca M. Possati

For a full list of titles in this series, please visit www.routledge.com/Antinomies/book-series/ANTIMN.

The Algorithmic Unconscious

How Psychoanalysis Helps in Understanding AI

Luca M. Possati

Routledge
Taylor & Francis Group
LONDON AND NEW YORK

First published 2021
by Routledge
2 Park Square, Milton Park, Abingdon, Oxon OX14 4RN

and by Routledge
52 Vanderbilt Avenue, New York, NY 10017

Routledge is an imprint of the Taylor & Francis Group, an informa business

British Library Cataloguing-in-Publication Data
A catalogue record for this book is available from the British Library

Library of Congress Cataloging-in-Publication Data
A catalog record has been requested for this book

ISBN: 978-0-367-69404-3 (hbk)
ISBN: 978-0-367-69405-0 (pbk)
ISBN: 978-1-003-14168-6 (ebk)

Typeset in Times New Roman
by Newgen Publishing UK

To E. & E.

Contents

Figures

Tables

Acknowledgments

I would like to express my gratitude to all the anonymous reviewers who helped me in improving my project and correcting my mistakes. I also thank the Institute of Philosophy of the University of Porto as well as the beautiful city of Porto, where I started thinking about this book. I especially thank Anthony Elliott, who believed in this project and accepted the book in the Routledge series Antinomies.

Mat Guzzo gave me tremendous support for the graphics of the book. I thank him very much.

The devil has a metal face.

(From *Man, Android and Machine*, by Phillip K. Dick)

Introduction

Beyond the metal face

"Do you know where you are?"

"I am in a dream."

"That's right, Dolores. You're in a dream. Would you like to wake up from this dream?"

"Yes. I'm terrified."

"There's nothing to be afraid of, Dolores, as long as you answer my questions correctly. Understand?"

"Yes."

Dolores and Bernard are sitting in front of each other. Their dialogue takes place in a dimly lit room in a surreal atmosphere. Dolores remembers her childhood and other fragments of her past, which was characterized by violence and pain. Dolores is a young girl, the daughter of a farmer named Peter, and she is in love with the unfortunate Teddy Flood. Bernard tries to help her remember, to understand what she feels, and to see whether inconsistencies or gaps exist in her memories. Both Dolores and Bernard are androids—artificial beings, or robots, that are identical to humans.

The dialogue between Dolores Abernathy and Bernard Lowe is the opening scene of the first season of *Westworld*, a series created by Jonathan Nolan and Lisa Joy and produced by HBO, which launched it in 2016. Dolores and Bernard are two of the main characters in a very complex story—inspired by a 1973 film written by Michael Crichton—which takes place in Westworld, a large, futuristic amusement park inspired by the Wild West and populated by androids that are identical to humans. These androids are humanoids, i.e., artificial duplicates, and are completely indistinguishable from humans. They have memories that programmers update continually. Only rich people can afford to visit the park, and visitors can indulge in their wildest fantasies without fear of retaliation by the hosts. The androids are, in fact, programmed to satisfy all human needs and are programmed not to harm humans in any way. Therefore, visitors are free from all social or moral inhibitions—they can even kill or rape robots without suffering any consequences. Even though they follow narratives that their programmers predefine, androids can vary the structure of these plots, thanks to their interaction with humans. The life of the androids is cyclical: at the end of each cycle, which often coincides with death, their memory is erased and reprogrammed.

The pivot of *Westworld*'s narrative mechanism is an accident: unbeknownst to the programmers at Delos (the company that manages the park), some androids have retained traces of memories from their previous lives or cycles due to a software bug. Thus, some androids begin to ask themselves questions about their identity, their existence, and the place where they live. They begin to develop authentic feelings, consciousness, and their own will. The appearance of these memories is what leads Dolores to embark on a journey to discover the "labyrinth," which she considers the key to understanding Westworld's origin. However, Dolores is not alone: Maeve Millay, another android, who is the matron of the park's brothel and is haunted by the memory of her daughter's brutal killing, decides to leave the park and triggers an android rebellion. Dolores sets off another rebellion inside the park by killing a human, Dr. Robert Ford, Westworld's co-founder and creative director.

Westworld is decidedly fictional as such perfect androids cannot exist today, but it also says something true about our society and our relationship with AI and digital technology. Today, this relationship is much more complex than before and poses a new series of problems in light of pervasive phenomena, such as the so-called "data deluge," the growing datafication of society, 3D printing, cloud computing, machine learning,[1] deep learning, and the development of platforms or video games—to mention just a few examples.

This book's central hypothesis is that the concepts and methods of psychoanalysis can be applied to the study of AI and human–AI interaction. This book is the first of its kind in that it aims to understand how psychoanalysis can be used to understand AI. The aim is not to apply AI to psychoanalysis, but rather to apply the concepts and methods of psychoanalysis to AI. It addresses philosophers of psychoanalysis and technology, psychoanalysts in the realm of AI and the applications of digital technology, and computer scientists (in the broadest sense of the term) in the fields of AI, philosophy, and psychoanalysis.

Several studies have proposed using AI to improve psychoanalysis and, in general, psychology. The predominant approach studies the transformations of personal identity through social networks, gaming, augmented reality, interactions with robotics, and simulation software. Turkle's works are the most popular example of such an approach. Turkle (1984, 1988) analyzes the psychological transformations that new technologies elicit in humans: "I look at the computer [...] not in terms of its nature as an analytical engine, but in terms of its second nature, as an evocative object, an object that fascinates, disturbs equanimity, and precipitates thought" (1984, 25). Technology

> catalyzes changes not only in what we do but in how we think. It changes people's awareness of themselves, of one another, of their relationship with the world. The new machine that stands behind the flashing digital signal, unlike the clock, the telescope, or the train, is a machine that "thinks." It challenges our notions not only of time and distance, but of mind.
>
> (34–5)

This line of research is developed through various works. For example, one (Turkle 2015) investigates the risks and opportunities of talking through digital technologies, while another (Turkle 2011) deals with the problem of the new solitude produced by a life continuously connected to the web. More technology means greater emotional solitude. Technology reshapes the emotional landscape of the self, which is divided dramatically between the screen and real, physical life, thereby yielding a fragmented, disoriented self.[2]

While Turkle moves from AI to psychology and psychoanalysis, I take the opposite approach: from psychoanalysis to AI. I want to underline two essential features of my approach: (a) the centrality of the concept of projective identification and (b) the reinterpretation of psychoanalysis in terms of actor-network theory. Furthermore, my approach is more philosophical than Turkle's. In some respects, I examine the possibilities of Turkle's inquiry. This does not mean that Turkle's investigations are unimportant to my research; indeed, her investigations confirm many of my hypotheses. She also speaks of "projection" of the self in digital technologies (i.e., the computer as a Rorschach test). For example, while analyzing the programming work of a group of children, Turkle distinguishes different programming styles that reflect personality and psychological dynamics: "The machine can act as a projection of part of the self, a mirror of the mind. The Rorschach provides ambiguous images onto which different forms can be projected. The computer, too, takes on many shapes and meanings" (Turkle 1984, 40); "[in using computers], the world the child creates, of rules or disorder, peace or violence, reflects back an image of who that child is or perhaps wants to be. Such images can help people move toward greater insight about conflicts, problems, and ways of thinking" (67). In this book, I try to develop this idea by applying it to AI, a sector of digital technology.

While Turkle mainly analyzes projection's effects on humans, I analyze the effects of projection on AI. In fact, I connect psychoanalysis and AI through the mediation of Latour's actor-network theory (ANT), which is a sociological model to analyze scientific facts and technology. This model's main characteristic is that of placing humans and non-humans on the same footing. In a nutshell, ANT "is a disparate family of material–semiotic tools, sensibilities, and methods of analysis that treat everything in the social and natural worlds as a continuously generated effect of the webs of relations within which they are located" (Law 2009, 141). ANT is not a theory; "theories usually try

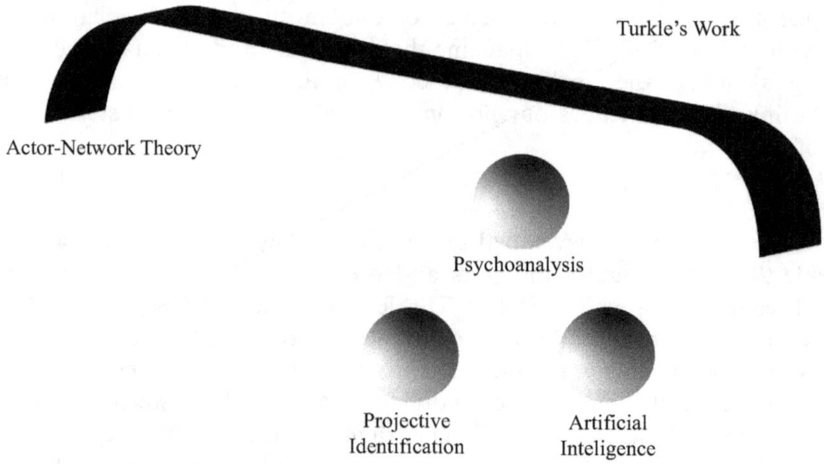

Figure I.1 The theoretical structure of this book.
The concept of projective identification plays a mediating role between psychoanalysis and AI.

to explain why something happens, but actor-network theory is descriptive rather than foundational in explanatory terms, which means that it is a disappointment for those seeking strong accounts" (Law 2009, 141).

This is an essential point for my inquiry (see Figure I.1). AI systems are conceived as social actors in the same context as that of humans. For this reason, I analyze both human unconscious dynamics and AI's technical aspects (e.g., miscomputation, noise in information, and algorithmic bias), which I link to the former. From this point of view, I use the psychoanalytic concept of projective identification, which is a powerful analytical and clinical tool. I apply projective identification to the study of AI. For instance, I claim that the concept of algorithmic bias is a kind of projective identification.

Many other works can be placed on the same methodological level with Turkle. Kunafo and LoBosco (2016) investigate the phenomenon of perversions in new technologies. Dean (2010) strives toward "thinking critically about media practices in a setting where they are fast, fun, and ubiquitous." Krzych (2010) studies the constitution of the post-human subject from a Lacanian perspective by analyzing, in particular, the aspects of Lacan's psychoanalysis as it relates to technology, particularly the concept of the "alethosphere." Developed during Seminar XVII, the "alethosphere" refers to an environment with gadgets that plug directly into human desires, and the notion is applied to a key technological, cultural, and political example (the interactive, touchscreen electoral maps made popular during the 2008 US election). Rambatan and Johanssen (2013) analyze cyborg culture in the context of digital technology (through the lens of the global warming, WikiLeaks, and Occupy movements). Bainbridge and Yates (2014) examine the possible applications of psychoanalysis to the study of new media and contemporary popular culture, an approach defined as "psycho-cultural" and which has its roots in object–relation psychoanalysis. Balik (2013) examines the unconscious dynamics behind the use of social networks and mobile phones. Singh (2014) interestingly analyzes, from a Jungian perspective, the TV series *Black Mirror* and the representation of the "always on" culture in it. Johanssen and Kruger (2016) question the way in which some psychoanalytic concepts can be applied to the study of digital media—a relevant contribution that is perhaps the largest of its kind.

Another group of studies partially related to Turkle examine the uses of AI from a clinical perspective. The main example of this approach is the work of Savege Scharff (2013, 2015), which concerns teleanalysis and teletherapy, i.e., psychotherapy by phone, Voice over Internet Protocol (VoIP), or videoteleconferencing (VTC). Scharff discusses the advantages and disadvantages of psychotherapy and psychoanalysis conducted over the phone and online. Other important studies of this type are Russell (2015) and Lemma and Caparotta (2014).

I also want to mention some texts that are very close to my approach. The first is Apprich (2018), which, starting from a critique of the physiological brain model in AI, develops a "psychoanalysis of things" and applies it to machine learning. This approach "may help to reveal some of the hidden layers within the current AI debate and hints toward a central mechanism in the psycho-economy of our socio-technological world." The second work, which is by Weinberg (1971), deals with the psychological aspects of programming. It is a fundamental book that opens up new perspectives on programming and computer science. This book wants, in part, to be an update of Weinberg's work. The third work, which is by Johanssen (2019), provides an introduction to Freud and Anzieu's psychoanalytic affect theories and applies them theoretically and methodologically to a number of case studies. Another very important, but more general, book is by Zwart (2019), who analyzes "technosciences" from a psychoanalytic and literary perspective. Its main sources are Lacan and Bachelard. Finally, I mention Nusselder (2009), which investigates the cyberspace through the core psychoanalytic notion of fantasy in Lacanian terms.

Can my approach be defined as post-phenomenological (e.g., Ihde 1979, 1990; Rosenberg and Verbeek 2017)? On the one hand, yes, because I analyze technologies in terms of the relationships between humans and technological artifacts, focusing on the different ways in which technologies shape the relationship between humans and the world. Furthermore, I try to combine philosophical and empirical approaches as much as possible. On the other hand, my method is not post-phenomenological for two reasons: first, because my investigation is not based on a critical dialogue with the phenomenological tradition, and second, because I believe that many post-phenomenologists make the mistake of talking about "technology" in general and forgetting that many different technologies exist and that each type poses different philosophical problems. Moreover, I think that the concept of mediation (the relationist approach) must be criticized as it is still based on an uncritical acceptance of Husserl's notion of intentionality (Rosenberg and Verbeek 2017, 24). From my perspective, the terms *relation* and *mediation* are much more obscure, complex, and stratified. Digital technology and AI are not only "mediators" between humans and the world. AI is more than a mediation: it is a completely new environment that redefines humans and their unconscious.

Psychoanalysis and AI: the odd couple

Psychoanalysis: a continuously expanding field

> From fin-de-siècle Vienna to the 21st century, psychoanalysis has been a powerful theoretical and therapeutic discourse for the critical analysis of our everyday life, interpersonal relationships, sociality, culture, politics, and history. Freud discovered the power of the "repressed unconscious" in the lives of women and men at the historical moment in which faith in science and its twin beacons of objectivity and rationality had moved center stage. While Freud powerfully deconstructed the age-old opposition between reason and passion, he nonetheless remained committed to the principle of scientific truth and the power of rationality to contribute to a better life, at once individual and collective. […] The unconscious is an otherness at the core of the self that can be profoundly troubling, but it is one that propels the self forward and drives culture and its various scientific enterprises onward. […] Freud characterized therapeutic action as a "talk-therapy" to better access the unconscious, to break down the repression barrier, thereby enabling the energy misdirected in the form of symptom formation to be rechanneled in (a) healthier direction.
>
> (Elliott and Prager 2019, 1)

Psychoanalysis is a continuously expanding field marked by rivalries between groups, theoretical debates, and transforming theories. Not a single psychoanalytic doctrine has been exempted from questioning and reformulation. Furthermore, psychoanalysis has been the subject of fierce attacks.[3]

This is certainly not the place to lay out a history of psychoanalysis, as it is impossible to summarize in a few lines all the developments of recent decades. The unity of psychoanalysis exists, but it is elusive and ever-changing. Much of the history of psychoanalysis after Freud is a critical revision of Freud.[4] For example, the US post-Freudian tradition and the British school of object relations have focused more on interpersonal and social aspects, rejecting Freudian metapsychology. In the US tradition, two directions dominate:

(a) ego psychology (Anna Freud, Hartmann, Kris, Lowenstein, Erikson, and Rapaport), and (b) the interpersonal and culturalist direction (Fromm, Sullivan, Thompson). "Ego psychology undoubtedly influenced psychoanalytic theory toward a deeper examination of interpersonal issues" (Elliott 2015, 27). The British school of object relations instead focuses more on the infantile dimension (Klein, Winnicott) and on the emotional relationship between the subject and his or her environment. The structure of the psyche is the result of emotional exchanges. Individuals seek a relationship not to satisfy their drives but for the goodness of the relationship itself. The autonomous self does not emerge against the backdrop of a transformation of desire; rather, it depends on the reconstruction of emotional links with other people.

The perspective of post-structuralist psychoanalysis, which is mainly French, is completely different. This approach accomplishes a deconstruction of the subject's autonomy and identity. The coherent and autonomous self is an illusion as the self is always influenced and conditioned by the unconscious. This idea is evident in the work of Lacan, where the ego is a narcissistic and paranoid illusion, and the psychic development of the subject consists precisely in freeing himself or herself from this illusion. Lacan sees the subject as repressively inscribed within a symbolic network of unstable signifiers. The illusion of the ego is at the heart—albeit in different ways—of Derrida's deconstructionist project, the writings of Lyotard and Althusser, and seminal books such as Deleuze and Guattari's *Anti-Oedipus* and *Mille Plateaux*.

Of course, these are just a few basic reference points. The key question is this: Is psychoanalysis still useful today? Certainly, "whatever the fluctuating stock-market fortunes of psychoanalysis in the area of mental and public health, Freud's impact has perhaps never been as far-reaching as it currently is within the public sphere and intellectual debate;" psychoanalysis today "is used by social and political theorists, literary and cultural critics, by feminists and postmodernists, such is its rich theoretical suggestiveness and powerful diagnosis of our current cultural malaise" (Elliott 2015, 5). This book is also a reflection on psychoanalysis today, on how it is being transformed by technology and AI. One of the book's central ideas is that the unconscious has a technological dimension, or vice versa. AI and digital technology give rise to a new dimension of the unconscious.

In the course of this book, I will, for the most part, consider Freudian and Lacanian psychoanalysis as my general framework.[5] This framework then will be enlarged and enriched through the integration of theories from Klein, Bion, Ogden, Solms, and neuropsychoanalysis. I approach psychoanalysis, and the study of the mind in general, from an anthropological and sociological perspective. As I have said above, the reference to Latour's works is essential to my research.[6] One of the central ideas of this book is that the unconscious must be viewed as a *collectif*, i.e., in Latourian terms, a network of human and non-human actors crisscrossed by relationships of force and transformations. The *collectif* is a theoretical model, not a metaphysical category: Latour's point of reference is always laboratory work, which was also the starting point for his work as a sociologist and anthropologist (Latour and Woolgar 1979). The *collectif* is not a thing, *res*, but a temporary process of association (Latour 2004, 202) in which humans and non-humans exchange properties and define each other to compose a temporary association, i.e., the *collectif* (Latour 2004, 97). My thesis is that AI is a development and a transformation of the *collectif* called the unconscious. Therefore, in my view, the *collectif* concept will mediate between the unconscious and AI.

I am aware of the numerous criticisms of Freudian and Lacanian ideas over the years (Van Rillaer 2019), but I believe that psychoanalytic theory's basic assertions remain valid. Nevertheless, my reading will not be neutral. I will provide an interpretation of these authors' concepts and theories from the perspective of the philosophy of technology. So far, the philosophy of technology has not dealt with psychoanalysis, just as psychoanalysis has not dealt with technology and the philosophy of technology. I intend to promote the interaction of these research fields. Psychoanalysis can provide us with a new perspective on technology in the same way that the philosophy of technology can open up a new perspective on psychoanalysis. Thanks to its relational and emotional dimensions, psychoanalysis prevents us from wasting time on empty discussions on "machine consciousness" and focuses directly on the human/machine relationship in a new way. I do not know what consciousness or intelligence is, and I have no desire to provide definitions. Is there a difference between "simulating a mind" and "literally having a mind"? I do not know. I cannot answer the question of whether machines in the future will have "consciousness" or "self-awareness" like that of humans. Instead, I know what behavior is: it is something empirical and observable. The behavior of humans and machines will, therefore, be at the center of my investigation.[7]

AI can be said in many ways

> We call ourselves *Homo sapiens*—man the wise—because our intelligence is so important to us. For thousands of years, we have tried to understand how we think; that is, how a mere handful of matter can perceive, understand, predict, and manipulate a world far larger and more complicated than itself. The field of artificial intelligence, or AI, goes further still: it attempts not just to understand but also to build intelligent entities. AI is one of the newest fields in science and engineering. Work started in earnest soon after World War II, and the name itself was coined in 1956. Along with molecular biology, AI is regularly cited as the "field I would most like to be in" by scientists in other disciplines. A student of physics might reasonably feel that all the good ideas have already been taken by

> Galileo, Newton, Einstein, and the rest. AI, on the other hand, still has openings for several full-time Einsteins and Edisons.
>
> (Russell and Norvig 2016, 1)

> Artificial Intelligence: The field of research concerned with making machines do things that people consider to require intelligence. There is no clear boundary between psychology and Artificial Intelligence because the brain itself is a kind of machine.
>
> (Minsky 1985, 326)

These general definitions of AI are taken from two of the most important books ever written on this topic: the first is from the textbook *Artificial Intelligence. A Modern Approach* by Stuart J. Russell and Peter Norvig; the second is from *The Society of Mind* by Marvin Minsky. Defining in precise terms the limits of AI is too complex a task to be completed in one book. The literature on this topic is vast and constantly growing. This book, therefore, does not aim to furnish a univocal definition, but to propose an analysis of AI from a psychoanalytic point of view. It is my deep conviction that the concept of AI is made up of three different, mutually influencing layers of meaning. The first layer is symbolic. Each symbol—as Ricoeur states (1965)—is a synthesis of archeology and teleology, in the sense that it tells us about the history of humanity and its future. AI symbolizes the way humans have reflected and reflect on their identity and history. For this reason, like any symbol, AI produces very powerful narratives and images: HAL 9000, Terminator, Blade Runner, Matrix, etc. The second layer of meaning is philosophical. AI contains an essentially philosophical problem, that of what the mind is and the relationship between mind and brain. The third layer is scientific and technological and is motivated by the attempt to answer, through machines, the questions posed by the previous two layers. As I said, these three levels are deeply connected, like the cycles of a spiral. This does not mean that the three levels are perfectly in sync; our imaginings about AI far outweigh its real possibilities, which often fail to recognize images or engage in even trivial conversations of the type humans can.

Paraphrasing Aristotle, AI is "said in many ways." In fact, the term *AI* can be used to refer to several phenomena: robotics, machine learning, deep neural networks, design, software, hardware, etc. The *Handbook of Artificial Intelligence* (Barr and Feigenbaum 2014) divides the AI field into seven main areas: knowledge representation, understanding natural language, learning, planning and problem-solving, inference, search, and vision. Nevertheless, the boundaries are constantly expanding. A few years ago, in a report on the state of AI research, a renowned scholar wrote that AI is "a branch of computer science that studies the properties of intelligence by synthesizing intelligence." The same report also acknowledged that "the lack of a precise, universally accepted definition of AI probably has helped the field to grow, blossom, and advance at an ever-accelerating pace" (*One Hundred Year Study on Artificial Intelligence*).[8] In sum, a single definition of AI does not exist.

From a technological point of view, AI (software systems, deep learning systems, advanced robotics, accelerating automata, and machine decision-making) is today a planetary phenomenon that is radically transforming our political and economic social systems and our relations with animals, our health, our safety, and our work (Colvin 2015, Ford 2015), as well as with ourselves and our personal identity. Our era is that of the "algorithmic war" (Amoore 2009), in which companies, institutions, and governments struggle daily to obtain and manage our data for security, commercial, scientific, or political propaganda purposes. We live in the age of "knowing capitalism" (Thrift 2005), i.e., capitalism that thrives on data because information generates profit. Therefore, capitalism is increasingly dependent on digital technologies and innovation. The data–information–capitalism circle produces new forms of social organization, power, and inequities (Baldwin 2016), knowing capitalism is a hyper-controlled form of society and economy in which privacy and security are jeopardized constantly. The concepts of reflexivity and life also undergo changes, as Giddens points out in his most recent works.[9] In short,

> technological innovation will very likely continue to open new paths into the global economy and continue to spawn uneven reversals and countertrends such as Brexit in the U.K., Trumpism in the U.S., and the rise of right-wing populism in Europe.
>
> (Elliott 2018, 70)

What characterizes AI is interaction with humans in everyday life. It transforms the entire human environment. A virtual assistant (e.g., Apple's Siri or Microsoft's Cortana) must be able to help us with what we need to do every day, actively collaborating with us and adapting to different fluid contexts. The WePod self-driving system must be able to transport passengers without crashing anywhere by adapting to different conditions on the road and in the surrounding environment, just like a human would. Active collaboration is also required of 3D printing technologies (Elliott 2018, 22) and robots in space exploration. An interesting example is Robonaut 2 (R2), a humanoid robot designed by NASA that can collaborate with astronauts: "Our challenge is to build machines that can help humans work and explore in space. Working side by side with humans, or going where the risks are too great for people, Robonauts will expand our ability for construction and discovery."[10] Another example is the IBM Cimon project that was launched in 2018. Cimon is a floating robot that uses Watson AI technology to assist astronauts on the space station. Cimon sees, listens, understands, speaks, and can analyze facial expressions.

However, AI is not just robotics. Today, the most advanced frontier concerns neuromorphic computing, that is, a form of computing and engineering based on the imitation of the human brain. This type of approach aims to strengthen

AI systems and make them more plastic, that is, more capable of adapting to ever new situations. "Neuromorphic computing develops at the intersection of diverse research disciplines, including computational neuroscience, machine learning, microelectronics, and computer architecture, among others."[11] An example of a neuromorphic chip is Intel's "Loihi," which includes a total of 130,000 neurons, each of which can communicate with thousands of others (see Davies et al. 2018). The chip is built to learn continuously in non-predefined and unsupervised contexts—just like the human brain.

Three pioneers of cyberspace—Wiener (2013), Turing (1950), and von Neumann (1958)—laid AI's modern theoretical foundations. Without this fundamental revolution in logic and technology, we would not be talking about AI today (see Davis 2000, Dyson 2012).

Scholars generally consider the 1956 workshop at Dartmouth College—organized by the mathematician John McCarthy, a collaborator of Minsky, Shannon, and Rochester—to be AI's official birthplace. McCarthy invented the term *artificial intelligence*, as he wanted to distinguish this field of study from cybernetics (Mitchell 2019, 17). In a presentation at the workshop, McCarthy stated that this project's origin lies in "the conjecture that every aspect of learning or any other feature of intelligence can be in principle so precisely described that a machine can be made to simulate it" (McCarthy 2006). The conference dealt with a series of topics that are still at the center of this field of study: natural language processing, neural networks, machine learning, abstract concepts and reasoning, and creativity. "We think that a significant advance can be made in one or more of these problems if a carefully selected group of scientists work on it together for a summer" (McCarthy 2006; Mitchell 2019, 18). The 1956 workshop, in which important scholars such as Allen Newell and Herbert Simon participated, opened many avenues and often sparked enormous expectations that were never met. After the conference, AI research took two paths: the biological one (the imitation of the structure of the human brain) and the practical one (the need to create programs and machines capable of doing things much better than humans).

Two fundamental paradigms can be distinguished in AI: symbolic and subsymbolic (or connectionist) (Mitchell 2019, 21) – in reality this is only a theoretical distinction: in most cases the paradigms coexist. The former conceives of AI as a process of combining symbols on the basis of fixed rules and operations, which is inspired by mathematical logic (e.g., General Problem Solver, created by Simon and Newell). These systems are transparent: Humans decide everything by establishing certain rules.

Subsymbolic systems are not transparent. This paradigm conceives of AI as a system that can learn how to accomplish a task from data. The first model of this way of thinking was the perceptron, invented by Frank Rosenblatt in the late 1950s. It is a mathematical model that describes neuron functioning in the human brain. Therefore, a program can simulate information processing in neurons:[12] A perceptron is a simple program that makes a yes-or-no (1 or 0) decision "based on whether the sum of its weighted inputs meets a threshold value" (Mitchell 2019, 25). You probably make some decisions like this in your life;

> for example, you might get input from several friends on how much they liked a particular movie, but you trust some of those friends' taste in movies more than others. If the total amount of "friend enthusiasm"—giving more weight to your more trusted friends—is high enough (i.e., greater than some unconscious threshold), you decide to go to the movie.
> (Mitchell 2019, 25–6)

While the symbolic AI paradigm is more capable of following procedures in a deductive and deterministic way, the subsymbolic one is, instead, more fluid, inductive, and able to perform tasks such as recognizing natural language, images, or objects. Basically, the perceptron knows the weights assigned to each input (a number) and its threshold (a number), and it acts (output) by following these relations. Therefore, we can gradually adapt the weights and the threshold to the type of task to be performed and to incoming data. Step by step, the perceptron can learn from data and accomplish the task without errors (Le Cun 2019, 141). The concept of machine learning (Apaydin 2016) and its different types (supervised and unsupervised) derives from this fundamental idea.

In the 1960s, the dominant AI paradigm was the symbolic one. In 1969, Minsky and Papert published a book in which they showed that the perceptron model had few applications and that the learning process was too long and complex (Minsky and Papert 1969; Nagy 1991). Thus, most funding was earmarked for symbolic AI projects. Over the past 20 years, this situation has changed, and the subsymbolic paradigm has become dominant. Today, the most used AI system involves so-called "neural networks," which are networks of artificial neurons. This technology began to develop in the 1980s from what is viewed as the "Bible" of AI connectionism, i.e., Rumelhart and McClelland (1986). Thanks to the Internet and the "big data deluge," we live in an "AI Spring" today: this quantum leap in machine intelligence "has emerged through exponentially growing quantities of data and processing power together with the development of complex algorithms, leading to new capacities for self-organization, sense-making, insight extracting, and problem solving" (Elliott 2018, 21). Advances in cloud computing, machine-to-machine communications, and the Internet of Things "have simultaneously developed incredibly rapidly, faster than previous technologies and with huge mobility consequences for how we live increasingly mobile lives as well as for enterprises and institutions" (Elliott 2018, 21).

The current AI paradigm is unsymbolic and connectionist. Neural networks do not proceed in a rigid, deductive way but in a fluid, inductive way, i.e., by analyzing probabilities in data (Kelleher and Tierney 2019). They grow and develop by processing data and recognizing recurring patterns in them.

Neural networks are learning systems modeled on the human brain (Bruder 2017). The core idea is to "mimic" the brain's structure in electronic form, whereby artificial neurons map their own connections during a self-learning process.[13] AI systems are intelligent to the extent that they can simulate human behavior: unlike the symbolic representation of the world, "the brain's learning power is by now simulated on digital computers, in order to automate the procedures by which a network of neurons learns to discriminate patterns and respond appropriately" (Apprich 2018, 29). The development of this new AI paradigm "can be explained by the concurrence of at least three mutually independent areas in the last years: deep learning, network analytics, and big data" (Apprich 2018, 31). Important convolutional networks—such as Detectron, Densepose, and Fairseq, developed by Facebook—are based on this learning paradigm.

A perfect example of the transition from the symbolic to the connectionist paradigm in AI is Google Translate. Since 2016—after five years of research on neural networks—the system has taken on a new configuration, one no longer based on the symbolic paradigm but on connectionism. The result has been a significant improvement in performance.

AI is an old dream of humanity (Riskin 2007; Wilson 2002). It also can be traced back to Aristotle, who spoke of slaves as machines (*Politics* 1, 4–5). There is a long history of thinking about humans and machines and artificial creatures. In Mary Shelley's *Frankenstein*, the scientist Victor Frankenstein creates a human-like being from the parts of inanimate bodies. Frankenstein creates an artificial life form thanks to his scientific and technical knowledge. Today, this dream has been translated into a new form—a digital one, i.e., the new language created by Turing and von Neumann. The key idea can be summarized with the definition provided by the UK government's 2017 "Industrial Strategy White Paper" (see Elliott 2018, 15), according to which, AI is technology "with the ability to perform tasks that would otherwise require human intelligence, such as visual perception, speech recognition, and language translation." This definition distinguishes intelligence from agency, i.e., from the ability to perform certain tasks that we view as essentially human. This is an important point: it is possible to accomplish tasks and achieve "intelligent" goals even if you are not intelligent or conscious as we understand them, i.e., if you are not human. This was confirmed by Searle (1980, 1999) through the so-called Chinese Room Argument. The man in the room does not understand a word of Chinese but has a program to use for interpreting questions that he is asked. A wide-ranging debate has surfaced on the subject, but I will not deal with it here. I think that it is useless to start from a univocal and rigid definition of *human intelligence* to define what is and what is not AI. We do not know what intelligence and consciousness are. We do not really know how our brain works.

One of the most important ideas in this book is that a correct approach to AI can only be empirical and "external," e.g., a machine can be called intelligent to the extent that it can simulate a human, namely, by performing operations that typically are executed by or acceptable to a human. As I asked above: Is there a difference between "simulating a mind" and "literally having a mind"? I do not know. I do not know what "having consciousness" means; therefore, I limit myself to what I see, namely, the interactions between humans and AI. The internal structure of a deep neural network is modeled on the human brain, but it also is essentially different. The machine's behavior, not its structure, is intelligent. I assert that *AI is any computational system that can interact—emit and react to stimuli actively and significantly—with a human and his or her environment.* This is the methodological approach that I call "machine behavior" in the first chapter of this book.

However, a rigidly behavioristic approach, completely reduced to the observable, represents harmful dogmatism. Psychoanalysis teaches that behavior is a text that must be interpreted. This is even more true for another type of AI that is much more complex than robots. Cimon and Robonaut are comparatively "simple" examples as we can see and interact with them. Much more complex is the case of the "complex digital systems" (Elliott 2018, 27) that act invisibly and manage most of the processes of our personal and economic social lives. These systems are ubiquitous, pervasive, always active, and manage billions of data per second. For example, think of algorithmic trading—huge computational systems that regulate investment banks, pension funds, mutual funds, and hedge funds. In this case, AI does not interact with the human environment but completely redefines it. AI "is not an advancement of technology, but rather the metamorphosis of all technology" (Elliott 2018, 5). A metaphor could be that of a "tsunami" that submerges everything. AI and all digital technologies are the "sea" in which humans and other previous technologies are immersed. This "sea" developed in the early part of the twentieth century. Today, the depths of this "sea" have become dark and impenetrable. The problem is that complex technological and social systems, including the conditions of systems reproduction,

> are characterized by unpredictability, non-linearity, and reversal. The ordering and reordering of systems, structures and networks, as developed in complexity theory, [are] highly dynamic, processual, and unpredictable; the impact of positive and negative feedback loops shifts systems away from states of equilibrium.
>
> (Elliott 2018, 26–7)

The human subjects who move in this new environment are not and cannot be fully aware of where (and who) they are. They are largely determined and conditioned. For this reason, simply observing the behavior of these systems could be useless. It must be also interpreted.

This book's scope and results

This book's central thesis is that AI's essence, i.e., what distinguishes it from any other technological form, is its unconscious roots. This does not mean that AI exerts a particular impact on the psychological dimension of humans. All technologies make an impact because they interact with the human environment and modify it in different ways. Instead, I argue that AI not only exerts a psychological impact but, above all, has psychological roots. As I said above, the central concept of my research is that of projective identification, which will be analyzed from the perspective of different therapists (Klein, Bion, Ogden, and Winnicott). Based on this concept, I speak of "emotional programming," which is this book's original thesis. The expression may shock many people (especially engineers, I imagine), but I believe that it is justified. One of Freud's fundamental teachings is that our emotions and fantasies are rooted in the past and define the backdrop of our present experiences. In any experience, we project emotional and imaginary content built from our past experiences. This also applies to technology and AI. As Turkle (1984, 34) says, "When you program it, it becomes your creature." To use Bion's metaphor, emotions and fantasies comprise the "theater" in which all other aspects—thoughts, desires, will, life and work projects, technologies—are called on "to play a role." This book aims to apply this metaphor to AI. A special type of unconscious projection takes place in AI that deeply characterizes the digital world's identity, social and otherwise. AI is not an "out-there" phenomenon. On the contrary, "the digital is both around us and inside us," and this implies that "robotics and AI become raw materials for the production and performance of the self" (Elliott 2018, 22). The construction of self passes through software and AI, and this process leaves deep traces in these technologies too.

This book is divided into five chapters. In Chapter 1, I define my approach to psychoanalysis and AI. The reader will not find a discussion about machine "consciousness." My approach is phenomenological and behavioral in nature. I do not presuppose rigid definitions of intelligence or consciousness as being applicable to humans and machines. AI will always remain a simulation of human abilities and attributes. In the literature, this is called the "machine behavior approach".

In Chapter 2, I pose two questions: Is psychoanalysis of artifacts possible? Does technique play a role in the formation of the unconscious? In doing so, I propose a reinterpretation of some fundamental Lacanian concepts through Latour's actor-network theory. I argue that this reinterpretation provides solid arguments in favor of applying psychoanalysis to AI.

In Chapter 3, I delve into questions regarding the algorithmic[14] unconscious, i.e., AI's unconscious roots. My thesis is that the human need for intelligent machines is rooted in the unconscious mechanism of projective identification, i.e., a form of emotional and imaginary exchange. Projective identification is an unconscious process of the imagination in which the ego projects parts of itself (qualities, abilities, or body parts) onto another person. This process is a form of unconscious communication: the projecting ego asks the person who receives the projected content to accept and contain it. I apply this dynamic to the sphere of artifacts. I do not wish to develop a "theory of affects." Instead, I argue that unconscious projective identification is a useful concept for analyzing and better understanding AI systems' behavior. Therefore, analyzing the projective identification processes that take place among groups of programmers and designers can be an important tool for understanding why AI systems behave in one way and not another. I assert that projective identification *is a form of emotional programming* that precedes all other forms of programming in AI. I see this as an original point that opens up new research possibilities in relation to the study of AI's emotional and affective dimensions—the emotional programming has to be thought *not as an effect but as a condition of the technical and engineering fact.* I clarify this point in Section 3.4, in which I explain what this emotional programming is and how it works in AI.

In Chapter 4, I analyze four concrete phenomena (errors, noise information, algorithmic bias, and AI sleeping) that I consider to be some of the most relevant expressions of the unconscious algorithmic, namely, the ways by which the work of projective identification appears, or "comes out," in AI. This serves the purpose of introducing the main result of the book: the topic of the algorithmic unconscious that is a theoretical model to study AI system's behavior. In the appendix of Chapter 4, I further expand my investigation by distinguishing another, even deeper, sense of the algorithmic unconscious, that is, the set of large software systems: a collection of billions of lines of code produced by thousands of programmers. These billions of lines of code that manage every activity in our lives exceed individual minds and consciousness.

In Chapter 5, I advance a new line of research. I claim that neuropsychoanalysis and the affective neurosciences can provide a new paradigm for AI research. So far, research in AI has always focused on the activities of the cerebral cortex (language, logic, memory, cognition, etc.). Now, the time has come to conceptualize and develop an AI of the subcortical. My hypothesis is that an artificial general intelligence (AGI) inspired by neuropsychoanalysis and affective neuroscience must be based on the simulation of the basic affective states of the human being. Here, I mainly refer to the work of Solms and Panksepp.

The algorithmic unconscious hypothesis explains the originality and complexity of AI compared to any other technological form. Starting from this hypothesis, I construct a topic that can be a useful theoretical model for analyzing AI systems, their behavior, and especially biases. Most noteworthy is the fact that this hypothesis can provide us with a new perspective on AI and

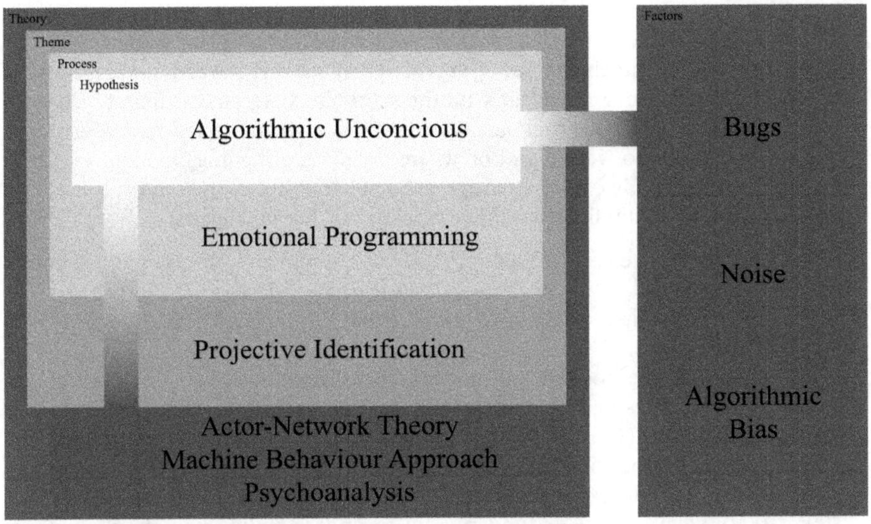

Figure I.2 The main topics of the book, and their relations.
Projective identification is an unconscious process of imagination and emotion. I claim that projective identification is a form of emotional programming that precedes all other forms of programming in AI. I analyze four concrete phenomena (errors, noise information, algorithmic bias, and AI sleeping) that I consider to be some of the most relevant expressions of the unconscious algorithmic. This serves to introduce the main result of the book: the topic of the algorithmic unconscious that is a theoretical model to study AI system's behavior.

intelligence. If we desire to create truly intelligent machines, we must create machines that are also "emotional," i.e., machines with the ability to welcome and deal with our emotions while possessing their own emotions as well. We cannot keep emotions, technology, and logic separate. Finally, it is not my goal to provide definitive answers. My intention in writing this book (for an overview, see Figure I.2) is to ask questions and open lines of inquiry; I do not claim to have defined a doctrine within these pages.

Notes

1 For a general definition of machine learning and related concepts, see developers.google.com/machine-learning/glossary/.

2 See also the criticism of Turkle's approach in Elliott (2018, 87–8):

> In Turkle's writings, the individual appears largely passive in relation to the digital world, internalizing demands for immediate emotional response and privileging the digital context. However, it is important to emphasize the contextual settings of technology. That is to say, it is crucial to grasp how digital materials are taken up, appropriated and responded to by individuals; it is

crucial to understand that people will respond in very different ways as they encounter new digital experiences, and importantly that their responses will change over time as they find new ways of responding to, and coping with, the opportunities and challenges of digital life. What is required is a focus on how people draw from and engage with the symbolic materials of digital technologies in order to reconstruct narratives of their lives and to reinvent versions of identity—of who we are and of where our lives are going. In sum, we need, among other things, to be attentive, as social analysts, to the opportunities and demands of digital life from the perspective of human agents.

3 This literature is extensive. I especially refer to Grünbaum (1984) and Obholzer (1982).

4 I refer here to principal classical works on Freud's life and doctrine: Jones (1953) and Sulloway (1979).

5 In the course of the book, I will not take into account either the development of Lacanian theories undertaken by authors such as Althusser or Castoriadis or Lacanian feminism (Kristeva, Irigaray, and Butler). I will not even consider the scholar who could be called the main, or the most popular, interpreter of Lacan today: Zizek. This does not mean that I am underestimating Zizek's work (Zizek 2006, 2009), which I view as fundamental to understanding how psychoanalysis can be applied to contemporary society and the analysis of ideology.

6 This means that I will not deal with more specific problems in this book, such as the mind–brain or mind–body relationship. I will express my position on these problems only in the fifth and last chapter, where I support a position inspired by neuropsychoanalysis and the affective neurosciences.

7 One of the main criticisms that can be advanced to the presentation and use of psychoanalysis in this book is its Eurocentric character. What do Freud's discoveries represent for those peoples who do not belong to European culture? Psychoanalysis can offer many theoretical tools for analyzing post-colonialism and racism, as the studies of Frosh (2010, 2013) and Khanna (2003) show. Hook (2012) developed an interesting analysis of apartheid through Lacan. Psychoanalysis is also widespread in Asia (Gerlach, Hooke, and Varvin 2018).

8 //ai100.stanford.edu/.

9 //player.fm/series/social-europe-podcast/how-the-digital-revolution-transforms-our-social-and-economic-lives.

10 //robonaut.jsc.nasa.gov/R2/.

11 intel.it/content/www/it/it/research/neuromorphic-computing.html.

12 A program essentially comprises three components: (a) the data structure and the operations connected to the data; (b) algorithms or finite procedures that transform certain incoming data into outgoing data; and (c) operations that maintain interaction with the surrounding environment (e.g., ensuring that the program behaves in the expected way). See Printz (2006, 32). For the program's chain assembly, see scheme 2.2 in Printz (2006, 35).

13 It is more accurate to say that AI is an abstract structure *inspired* by the brain, but it does not copy it.

14 The notion of an algorithm is complex and multifaceted; thus, it is almost impossible to provide a single definition. See Primiero (2020, chap. 6). More on this later.

References

Amoore, L. 2009. "Algorithmic War: Everyday Geographies of the War on Terror." *Antipode* 41, no. 1: 49–69.

Apaydin, E. 2016. *Machine Learning: The New AI.* Cambridge, MA: MIT Press.

Apprich, C. 2018. "Secret Agents. A Psychoanalytic Critique of Artificial Intelligence and Machine Learning." *Digital Culture and Society* 4, no. 1: 30–44.

Bainbridge, C., and C. Yates, eds. 2014. *Media and the Inner-World: Psycho-Cultural Approaches to Emotion, Media and Popular Culture.* London/New York: Palgrave MacMillan.

Baldwin, R. E. 2016. *The Great Convergence: Information Technology and the New Globalization.* Cambridge, MA: Harvard University Press.

Balik, A. 2013. *The Psychodynamics of Social Networking.* London/New York: Routledge.

Barr, A., and E. A. Feigenbaum. 2014. *The Handbook of Artificial Intelligence.* Oxford: Butterworth-Heinemann.

Bruder, J. 2017. "Infrastructural Intelligence: Contemporary Entanglements between Neuroscience and AI." *Progress in Brain Research* 233: 101–28.

Colvin, G. 2015. *Humans Are Underrated: What High Achievers Know that Brilliant Machines Never Will.* New York: Penguin.

Davis, M. 2000. *The Universal Computer. The Road from Leibniz to Turing.* New York: W. W. Norton.

Davies, M., N. Srinivasa, T.-H. Lin, G. N. Chinya, Y. Cao, S. H. Choday, G. D. Dimou, P. Joshi, N. Imam, S. Jain, Y. Liao, C.-K. Lin, A. Lines, R. Liu, D. Mathaikutty, S. McCoy, A. Paul, J. Tse, G. Venkataramanan, Y.-H. Weng, A. Wild, Y. Yang, and H. Wang. 2018. "Loihi: A Neuromorphic Manycore Processor with On-Chip Learning." *IEEE Micro* 38, no. 1: 82–99.

Dean, J. 2010. *Blog Theory: Feedback and Capture in the Circuits of Drive.* Cambridge, MA: Polity.

Dyson, G. 2012. *Turing's Cathedral. The Origins of the Digital Universe.* London: Penguin Books.

Elliott, A. 2015. *Psychoanalytic Theory.* London/New York: Palgrave Macmillan.

———. 2018. *The Culture of AI: Everyday Life and the Digital Revolution.* London/New York: Routledge.

Elliott, A., and J. Prager, eds. 2019. *The* Routledge *Handbook of Psychoanalysis in the Social Sciences and Humanities.* London/New York: Routledge.

Ford, M. 2015. *The Rise of the Robots: Technology and the Threat of a Jobless Future.* New York: Basic Books.

Frosh, S. 2010. *Psychoanalysis Outside the Clinic: Interventions in Psychosocial Studies.* London/New York: Red Globe Press.

———. 2013. *Hauntings: Psychoanalysis and Ghostly Transformations.* London/New York: Palgrave Macmillan.

Gerlach, A., M. T. Hooke, and S. Varvin, eds. 2018. *Psychoanalysis in Asia.* London/New York: Routledge.

Grünbaum, A. 1984. *The Foundations of Psychoanalysis. A Philosophical Critique.* Berkeley/Los Angeles/London: University of California.

Hook, D. 2012. *The Mind of Apartheid. A Critical Psychology of Postcolonial.* London/New York: Routledge.

Ihde, D. 1979. *Technics and Praxis. A Philosophy of Technology*. Berlin: Springer.
———. 1990. *Technology and Lifeworld. From Garden to Earth*. Bloomington: Indiana University Press.
Johanssen, J. 2019. *Psychoanalysis and Digital Culture*. London/New York: Routledge.
Johanssen, J., and S. Krüger. 2016. *Digital Media, Psychoanalysis and the Subject*. London/New York: Palgrave McMillan.
Jones, E. 1953. *The Life and the Work of Sigmund Freud*. New York: Basic Books.
Kelleher, J., and B. Tierney. 2019. *Data Science*. Cambridge, MA: MIT Press.
Khanna, R. 2003. *Dark Continents: Psychoanalysis and Colonialism*. Durham, NC: Duke University Press.
Krzych, S. 2010. "Phatic Touch, or the Instance of the Gadget in the Unconscious." *Paragraph* 33, no. 3: 379–91.
Kunafo, D., and R. LoBosco, eds. 2016. *The Age of Perversion. Desire and Technology in Psychoanalysis and Culture*. Abingdon: Taylor & Francis.
Latour, B. 2004. *Politiques de la nature*. Paris: La Découverte.
Latour, B., and S. Woolgar. 1979. *Laboratory Life. The Construction of Scientific Facts*. Thousand Oaks, CA: Sage Publications.
Law, J. 2009. "Actor-Network Theory and Material Semiotics." In *The New Blackwell Companion to Social Theory*, edited by S. Turner Bryan, 141–58. Oxford: Blackwell.
Le Cun, Y. 2019. *Quand la machine apprend. La revolution des neurons artificiels et de l'apprentissage profond*. Paris: Odile Jacob.
Lemma, A., and L. Caparrotta, eds. 2014. *Psychoanalysis in the Technoculture Era*. London: Routledge.
McCarthy, J. 2006. "A Proposal for the Dartmouth Summer Research Project in Artificial Intelligence." Submitted to the Rockefeller Foundation, 1955. Reprinted in AI Magazine 27, no. 4: 12–4.
Minsky, M. 1985. *The Society of Mind*. New York: Simon & Schuster.
Minsky, M., and S. Papert. 1969. *Perceptrons: An Introduction to Computational Geometry*. Cambridge, MA: MIT Press.
Mitchell, M. 2019. *Artificial Intelligence. A Guide for Thinking Humans*. New York: Farrar, Strauss and Giroux.
Nagy, G. 1991. "Neural Networks—Then and Now." *IEEE Transactions on Neural Networks* 2, no. 2: 316–8.
Nusselder, A. 2009. *Interface Fantasy. A Lacanian Cyborg Ontology*. Cambridge, MA: MIT Press.
Obholzer, K. 1982. *The Wolf-Man Sixty Years Later: Conversations with Freud's Controversial Patient*. London: Routledge & Kegan Paul.
Primiero, G. 2020. *On the Foundations of Computing*. Oxford: Oxford University Press.
Printz, J. 2006. *Architecture Logicielle*. Paris: Dunod.
Rambatan, B., and J. Johanssen. 2013. *Cyborg Subjects: Discourses on Digital Culture*. Seattle: CreateSpace Publishing.
Ricoeur, P. 1965. *De l'interprétation. Essai sur Freud*. Paris: Seuil.
Riskin, J., ed. 2007. *Genesis Redux: Essays in the History and Philosophy of Artificial Life*. Chicago: University of Chicago Press.
Rosenberg, R., and P.-P Verbeek. 2017. *Postphenomenological Investigations*. London: Routledge.
Rumelhart, D., and J. McClelland. 1986. *Parallel Distributed Processing: Explorations in the Microstructure of Cognition*. Cambridge, MA: MIT Press.

Russell, G. 2015. *Screen Relations: The Limits of Computer-Mediated Psychoanalysis and Psychotherapy*. London/New York: Routledge.
Russell, S., and P. Norvig. 2016. *Artificial Intelligence. A Modern Approach*. London: Pearson.
Savege Scharff, J. 2013. "Technology-Assisted Psychoanalysis." *Journal of the American Psychoanalytic Association* 51: 101–30.
———, ed. 2015. *Psychoanalysis On-line*. London/New York: Routledge.
Searle, J. 1980. "Minds, Brains, and Programs." *Behavioral and Brain Sciences* 3, no. 3: 417–57.
———. 1999. "The Chinese Room." In *The MIT Encyclopedia of the Cognitive Sciences*, edited by R. A. Wilson and F. Keil. Cambridge, MA: MIT Press.
Singh, G. 2014. "Recognition and the Image of Mastery as Themes in Black Mirror. An Eco-Jungian Approach to 'Always on Culture.'" *Culture. International Journal of Jungian Studies* 6 no. 29: 120–32.
Sulloway, F. 1979. *Freud, Biologist of the Mind*. New York: Burnett.
Thrift, N. 2005. *Knowing Capitalism*. New York: Sage Publications.
Turing, A. 1950. "Computing Machinery and Intelligence." *Mind* 49: 433–60.
Turkle, S. 1984. *The Second Self: Computer and the Human Spirit*. Cambridge, MA: MIT Press.
———. 1988. "Artificial Intelligence and Psychoanalysis: A New Alliance." *Daedalus* 117, no. 1: 241–68.
———. 2011. *Alone Together*. New York: Basic Books.
———. 2015. *Reclaiming Conversation: The Power of Talk in a Digital Age*. New York: Penguin Press.
Van Rillaer, J. 2019. *Freud & Lacan. Des charlatans?* Brussels: Mardaga.
Von Neumann, J. 1958. *The Computer and the Brain*. New Haven/London: Yale University Press.
Weinberg, G. 1971. *The Psychology of Computer Programming*. New York: Van Nostrand Reinhold Company.
Wiener, N. 2013. *Cybernetics, or Control and Communication in the Animal and the Machine*. Eastford: Martino Fine Books.
Wilson, T. D. 2002. *Strangers to Ourselves. Discovering the Adaptive Unconscious*. Cambridge, MA: Harvard University Press.
Zizek, S. 2006. *How to Read Lacan*. London: Granta.
———. 2009. *The Sublime Object of Ideology*. London/New York: Verso Books.
Zwart, H. 2019. *Psychoanalysis of Technoscience: Symbolisation and Imagination*. Münster: LIT Verlag.

Chapter 1

Why an algorithmic unconscious for AI?

1.1 What this book is about

Positing the existence of an "unconscious" of AI systems[1] and working scientifically on the basis of this hypothesis are controversial. Why are we forced to acknowledge an "unconscious" for this type of machine? In what sense do we speak of an "unconscious" in relation to AI? What does this hypothesis add to our understanding of AI? In this book, I argue that the hypothesis of an AI unconscious, or an "algorithmic unconscious," is legitimate and necessary.

The hypothesis of an algorithmic unconscious is *legitimate* because it is perfectly in line with psychoanalysis and the anthropology of science (see Chapter 2). We can develop a coherent theory that allows us to extend the Freudian notion of the unconscious to AI.

This hypothesis is also *necessary*. A basic fact has to be considered here: an intelligent machine cannot be an imitation of the brain. Our knowledge of how the human brain works is very limited. Thus, how can we imitate something we do not understand? The analogy between the brain and AI is not only limited—it is also misleading. A machine can be said to be intelligent only by the interaction we have with it, i.e., from the ability that the machine has to collaborate with us in a useful way.

Machines and humans are completely different beings. When we talk about "artificial intelligence," we are talking about the interaction between machines and human beings. AI is, therefore, the name of an intermediate field between machines and humans. It is a multifaceted concept that takes on different meanings in relation to different contexts. "To warrant the label of an intelligent system, an IT system must collaborate, and for this, it must follow [human] reasoning step by step, be careful, and know exactly where we are in the path that leads to the solution" (Jorion [1989] 2012, 20; my translation). An IT system can demonstrate this ability "by offering *the most relevant information* to its user" (20; my translation). The IT system is not a system that just repeats; it knows how to learn from humans and, eventually, corrects them. "Imitation" means here "negotiation." "In order to be worthy of being called intelligent, an IT system must not only demonstrate its knowledge but must also know how to negotiate with its user" (Jorion [1989] 2012, 21; my translation).

AI is a hermeneutic space. This implies that AI is always the result of an act of interpretation. When we say that a machine is intelligent, we start with an interpretation of its behavior in relation to ours. For this reason, I view the application of psychoanalysis to the study of AI as necessary. Psychoanalysis is, above all, an extraordinary tool for interpreting human behavior and its relationship with the environment. Moreover, many AI systems have a computational structure that is too extensive and complex to be fully understood. A non-technical approach is, therefore, essential to understanding how AI acts and why it acts in the way it does. As I will show in the next section, my approach is focused on the notion of "machine behavior," which is empirical, phenomenological, and also hermeneutical, together forming the methodological perspective of the present book, which is, therefore, not a rigid form of behaviorism that reduces everything to the observable.

Neuroscience and economics have shown that emotions play a fundamental role in rational thinking and decision-making (Damasio 1994, 1999; Kahneman 2011). Psychoanalysis makes the same claim, although from a different point of view. Psychoanalysis holds that intelligence derives from the emotions and the ability to understand and manage them. However, it does not reduce emotions to simple physiological mechanisms. For psychoanalysis, emotions arise and develop in a relational context and have an important organizational function in relation to memory and cognition. An artificial neural network constantly interacts with its environment, but this interaction is considered only from the point of view of probability and data strings of 1 and 0. The hypothesis of this book is that there is another dimension of this interaction that is yet to be taken into consideration, i.e., the emotional aspect. I will argue that there is an emotional relation between AI systems and humans that cannot be translated into data; nevertheless, it is fundamental in order to understand AI behavior. Emotional dynamics organize data—in the sense that they are able to determine the relationships between different types of data (numeric and non-numeric). Without this type of organization, a machine can never be truly intelligent. If we do not fully understand the emotional side of AI, which is both human and artificial, we will never truly understand what an intelligent machine is; consequently, we will never be able to design and build machines that are truly intelligent. By the "emotional side of AI," I do not mean affective computing (Picard 1997), that is, the possibility of a computational system to simulate human emotions. I mean the relationship between the creator (or the user) and the machine. I will deal with affective computing in Chapter 5 from the perspective of neuropsychoanalysis and affective neuroscience.

Therefore, I consider the hypothesis of an "unconscious for AI" *necessary* for two reasons: (a) AI is a space of interpretation between humans and machines, in which the point of reference is the human intelligence; and (b) in human intelligence, emotion and feelings play an essential role, and psychoanalysis has investigated this aspect a lot.

Before developing these claims, two clarifications are warranted. The first concerns our methodological approach. Hypothesizing an "algorithmic unconscious" does not mean attributing consciousness, feelings, and impulses to AI. The philosophical debate about the consciousness of machines is very broad and cannot be summarized in a few paragraphs (for an overview, see Churchland 1990; Cummins and Cummins 1999; Larrey 2019). It is not my purpose here to formulate theses on the presumed "consciousness of machines." In order to do this, I would first need to define what I mean by "consciousness," "intention," and "intelligence" and then show how these concepts can be applied to AI. However, this is not my purpose here. What really interests me is the relationship between the human unconscious and AI and, in this context, how and why AI takes root in the human unconscious and develops it. I hold that AI is a *post-human unconscious*. AI shapes the human unconscious and produces a new form of unconscious that has yet to be explored. We need new tools in order to understand and explore this unconscious and, therefore, to truly understand who we have become.

The second clarification concerns the status of the notion of the unconscious. Freud's (1984) unconscious is not only the preconscious or the perceptive unconscious; it is not simply what does not reach consciousness. It is mainly that which is repressed, i.e., what *must not* become conscious. The concept of repression is absolutely fundamental to psychoanalysis (see Sulloway 1979, 64–7; see also Ellenberger 1970; Freud [1899] 1997; Elliott 2015).

To understand what repression is, we must first clarify a crucial point: Freud thinks of the mind as a system composed of several types of memory. This can be clearly seen in the letters to Fliess (6 December 1896) and in the seventh chapter of the *Interpretation of Dreams*. Memory is not a static deposit of traces. The fundamental mechanism of memory is recording, rewriting, and reorganization. Moreover, each type of memory has its own criteria. Psychic material is translated from one memory system to another, following the evolutionary development of the individual; in the course of this development, the psychic material must be rewritten and translated from one system to another. The memory traces written afterward influence the previous ones and their interpretation. Repression represents the failure of this process of transcription and translation; some material has not been translated and, therefore, continues to exist in a new system but with the rules of a previous system. The mind is unable to translate psychic material from one memory system to another. In other words, to implement its fundamental purposes, which are organization and control, the ego must make a choice and, consequently, decide not to translate psychic material that could be harmful and prevent its essential task: the mediation between internal drives and reality. The repressed is not canceled but remains fixed in a different, older, procedural and automatic memory—a type of memory which is very different from what is called working memory, that is, the ego, which, according to some neuroscientists, resides in the prefrontal cortex (Solms 1996). For this reason, the repressed returns, but it returns as repetition and as automation, as happens in transference. Repression is accomplished in the working memory, which selects the information coming from previous systems and decides what needs to be translated or not (Solms 2008). We need a specific technique, i.e., psychoanalysis, in order to analyze the repressed. According to Freud, unconscious thoughts evade psychic censorship through specific mechanisms, such as condensation (*Verdichtung*) and displacement (*Verschiebung*) (Freud [1899] 1997; Lacan 1953–54, 81–166; Ricoeur 1965, part 2).

This leads us to a second essential point, to which we will return later. The fundamental principle of psychoanalysis establishes not only that unconscious thoughts exist, but that *the mind is itself unconscious*. In Freud's psychic model, consciousness is a fragment whose ontological basis resides in the unconscious substratum. A continuous unconscious stream of uninterrupted mental events follows its own logic, which may be discerned by episodic manifestations in conscious thoughts, emotions, and behaviors; consciousness represents punctuated islands in this stream. Current research in cognitive science and neuroscience confirm basic Freudian insights: (a) certain cognitive processes are not only hidden but cannot be brought into consciousness; and (b) the self is fundamentally non-unified, "and because of its fragmentation, self-awareness represents only a small segment of cognitive processes" (Tauber 2014, 233). The existence of the unconscious has been conclusively demonstrated (see Wilson 2002).

The basis of my arguments in this book is Freud's "incompleteness theorem," according to which "extra-logical (largely unconscious) factors play a constitutive role in evaluating properties," and "the subordination of consciousness to unconscious mental processing also occurs in the setting of complex reasoning" (Tauber 2014, 235). With the greater capacity to process information, "unconscious thought processes [...] are better able to weigh alternative choices as well as to employ divergent thinking (as opposed to the convergence strategies of conscious thought)" (Tauber 2014, 235; see also Bos, Dijksterhuis, and Baaren 2008).

When I speak of an "algorithmic unconscious" in this book, I am not referring to the phenomenological notion of the preconscious but to Freud's "incompleteness theorem." However, in so doing, I also follow Lacan's interpretation. Lacan introduced another important element—language—and formulated an essential question: "Who speaks?" For our purposes, this question needs to be reformulated: "Who speaks in algorithms?"

1.2 Machine behavior: research perspectives and the status of the art

> [*W*]*e describe the emergence of an interdisciplinary field of scientific study*. This field is concerned with the scientific study of intelligent machines, not as engineering artefacts but as a class of actors with particular behavioral patterns and ecology. This field overlaps with, but is distinct from, computer science and robotics. It treats machine behavior empirically. This is akin to how ethology and behavioral ecology study animal behavior by integrating physiology and biochemistry—intrinsic properties—with the study of ecology and evolution—properties shaped by the environment. Animal and human behaviors cannot be fully understood without the study of the contexts in which behaviors occur. Machine behavior, similarly, cannot be fully understood without the integrated study of algorithms and the social environments in which algorithms operate. [...] Commentators and scholars from diverse fields—including, but not limited to, cognitive systems engineering, human computer interaction, human factors, science, technology and society, and safety engineering—are raising the alarm about the broad, unintended consequences of AI agents that can exhibit behaviors and produce downstream societal effects (both positive and negative) that are unanticipated by their creators.
>
> (Rahwan et al. 2019, 477; emphasis added)

The behavior of AI systems is often studied in a strict technical engineering and instrumental manner. Many scholars are usually only interested in what a machine does and the results it achieves. However, another—broader and richer—approach is possible, one that takes into account not only the purposes for which a machine is created and its performance but also its "life," i.e., its behavior as an agent that interacts with the surrounding environment (which includes humans and non-humans). The notion of AI behavior is entirely legitimate: it is a new form of behavior of which the study calls for new tools. AI is characterized by a particular kind of behavior that is original in nature and has its own structures and dynamics. This approach takes as its object "machine behavior," i.e., the study of AI behavior, "especially the behavior of black box algorithms in real-world settings" (Rahwan et al. 2019, 477) through the conceptual schemes and methods of the social sciences. "Rather than using metrics in the service of optimization against benchmarks, scholars of machine behavior *are interested in a broader set of indicators, much as social scientists explore a wide range of human behaviors in the realm of social, political or economic interactions*" (Rahwan et al. 2019, 479; emphasis added).

As various scholars claim, current algorithms are perfectly capable of adapting to new situations, even creating new forms of behavior. For instance, Cully et al. (2015) demonstrate that it is possible to construct an intelligent trial-and-error algorithm that "allows robots to adapt to damage in less than two minutes in large search spaces without requiring self-diagnosis or prespecified contingency plans" (503). When the robot is damaged, it uses prior knowledge "to guide a trial-and-error learning algorithm that conducts intelligent experiments to rapidly discover a behavior that compensates for the damage" (503), which enables the robot to adapt itself to many different possible situations, similar to animals. "This new algorithm will enable more robust, effective, autonomous robots, and may shed light on the principles that animals use to adapt to injury" (503). Lipson (2019) has obtained the same results with another robotic experiment about autonomy and robots. AI systems are capable of creating a completely new form of behavior by adapting themselves to new contexts, which calls attention to the notion of artificial creativity.

According to Rahwan et al. (2019), there are three fundamental reasons that make the notion of "machine behavior" inevitable:

> First, various kinds of algorithms operate in our society, and algorithms have an ever-increasing role in our daily activities. Second, because of the complex properties of these algorithms and the environments in which they operate, some of their attributes and behaviors can be difficult or impossible to formalize analytically. Third, because of their ubiquity and complexity, predicting the effects of intelligent algorithms on humanity—whether positive or negative—poses a substantial challenge.
>
> (478)

Studying the adaptation of AI to the environment also means studying the lack of this adaptation, i.e., the pathological behaviors that AI can develop. According to O'Neil (2016), the uncritical use of algorithms and AI systems can result in very dangerous consequences for society. Studying AI systems that process big data only from a mathematical and statistical point of view significantly undermines our understanding of the complexity of their functioning, hindering us from grasping the real issues that they imply. These AI systems can produce injustices, inequalities, and misunderstandings, feed prejudices and forms of discrimination and ethical and social problems, aggravate critical situations, or even create new ones. Furthermore, these systems are "black boxes," i.e., they are opaque. There are two explanations for this: (a) for legal and political reasons, their functioning is often not made accessible by the companies that create and use them; (b) the computation speed makes it impossible to understand not only the overall dynamics of the calculation but also the decisions that the systems make. Engineers struggle to explain why a certain algorithm has taken that action or how it will behave in another situation, e.g., in contact with other kinds of data (Voosen 2017, 22). These algorithms are complex and ubiquitous, and it is very difficult to predict what they will do in the most diverse contexts.

A typical example is that of college loans (O'Neil 2016, 81). The AI system identifies the poorest segment of the population and bombards them with ads for university loans, spreading the message—at least theoretically correct—that better education favors better employment and income. As a result, many poor people decide to take on debt. Unfortunately, due to periods of economic recession, people who have amassed debts to undergo training are unable to find jobs, or they lose what they already have, a situation that the mathematical model had not foreseen. The result is that people cannot repay their loans, leading to a situation in which the poor become poorer: the starting situation is amplified and made worse. AI systems behave in a pathological manner, i.e., they are unable to adapt to the human environment. Given the large amount of data that AI systems manage, they can create major problems.

A point of clarification is that the mathematical model is not always vitiated by a partial or inaccurate view of reality. An even more complex case is that of AI systems that assimilate models of thought or behavior from the data and then assume biases that modify their original behavior. Such systems can also be built on "good" models motivated by the best of intentions, but they can also learn from data to behave differently and inflict damage this is the case of the "algorithmic biases" that we will explore more in Chapter 4.

If we want to minimize the collateral effects of AI behavior, we must look beyond its logical–mathematical structure and statistics. We must choose a model of analysis that is not based on the rigid distinctions between humans and machines; instead, it should focus on their mutual interaction. Algorithms are not only simple abstract mathematical machines. Immersed in a world of action and in contact with human lives, an algorithm becomes an agent like any other, capable of both imitating and assimilating the behavior of other agents (human, animals, or other machines) or even acting in a new way.

Studying AI behavior is not at all easy. AI behavior can be analyzed from at least three perspectives: (a) the behavior of a single AI system; (b) the behavior of several AI systems that interact (without considering humans); and (c) the interaction between AI systems and humans. Today, most interactions on our planet are of type b. For Rahwan et al. (2019), when we talk of interactions between AI systems and humans, we mean three things: (c.1) how an AI system influences the behavior of humans; (c.2) how humans influence the behavior of AI systems; and (c.3) how humans and AI systems are connected within complex hybrid systems and, thus, collaborate, compete, or coordinate. All these layers are intertwined and influence each other, as shown in Figure 1.1.

The machine behavior approach underscores a fundamental point, which is the premise of the next chapter. Today, we can no longer study humans and machines as separate subjects. They constantly interact in a single ecosystem in which agency is distributed between humans and non-humans. Hence, we need an integrated approach. We cannot exclusively consider the consequences that machines have on humans. Today, as I said in the introduction, digital technology is an environment, an ecosystem that redefines

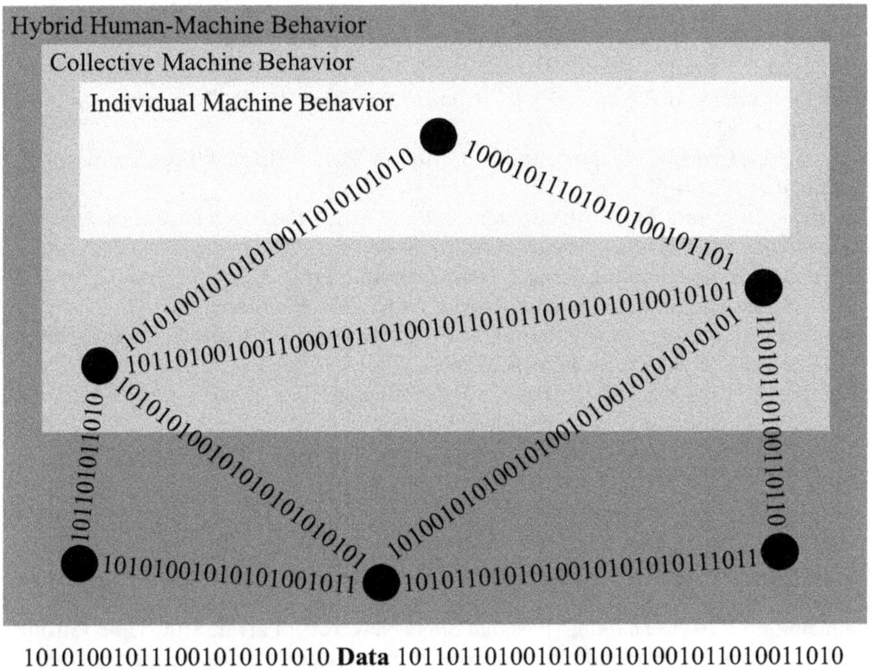

Figure 1.1 AI behavior and its contexts.
AI behavior can be analyzed from at least three perspectives: (a) the behavior of a single AI system; (b) the behavior of several AI systems that interact (without considering humans); and (c) the interaction between AI systems and humans.

everything inside it. We must reach a point of equilibrium beyond human/machine dualism. In the next chapters, I will try to apply methods and concepts from psychoanalysis to the machine behavior hypothesis. With this, however, I do not make the claim that psychoanalysis is a behaviorist science. Obviously, it is not.

Note

1 The literature on AI is extensive. I mainly refer to Le Cun (2019), Mitchell (2019), Davenport et al. (2019), and Agrawal et al. (2018). I also refer to some classic texts, such as Black (1988), Frost (1986), Rausch-Hindin (1988), and Winston and Prendergast (1984).

References

Agrawal, A., J. Gans, and A. Goldfarb. 2018. *Prediction Machines*. Boston, MA: Harvard Business Review Press.

Black, W. J. 1988. *Les systèmes intelligents basés sur la connaissance*. Paris: Masson.

Bos, M. W., Dijksterhuis, A., and R. B. van Baaren. 2008. "On the Goal Dependency of Unconscious Thought." *Journal of Experimental and Social Psychology* 44, no. 4: 1114–20.

Churchland, P., and P. Smith. 1990. "Could a Machine Think?" *Scientific American* 262, no. 1: 32–7.

Cully, A., J. Clune, D. Tarapore, and J.-B. Mouret. 2015. "Robots that Can Adapt like Animals." *Nature* 521: 503–7.

Cummins, R., and D. D. Cummins 1999. *Minds, Brains, and Computers. The Foundations of Cognitive Science*. Oxford: Wiley.

Damasio, A. 1994. *Descartes' Error*. New York: Putnam.

———. 1999. *The Strange Order of Things*. New York: Pantheon.

Davenport, T. H., E. Brynjolfsson, and A. McAfee. 2019. *Artificial Intelligence*. Boston, MA: Harvard Business Review.

Ellenberger, H. 1970. *The Discovery of Unconscious: The History and Evolution of Dynamic Psychiatry*. New York: Basic Books.

Elliott, A. 2015. *Psychoanalytic Theory*. London/New York: Palgrave Macmillan.

Freud, S. 1984. *On Metapsychology. The Theory of Psychoanalysis*. London: Penguin.

———. (1899) 1997. *The Interpretation of Dreams*. Reprint, Hertfordshire: Wordsworth.

Frost, R. A. 1986. *Introduction to Knowledge Base Systems*. London: Collins.

Jorion, P. (1989) 2012. *Principies des systems intelligentes*. Broissieux Bellecombe-en-Bauges: Ed. Croquant.

Kahneman, D. 2011. *Thinking, Fast and Slow*. New York: Farrar, Straus and Giroux.

Lacan, J. 1953–54. "Fonction et champ de la parole et du langage en psychalyse." In *La Psychanalyse*. Paris: PUF.

Larrey, P. 2019. *Artificial Humanity. An Essay on the Philosophy of Artificial Intelligence*. Rome: IFPress.

Le Cun, Y. 2019. *Quand la machine apprend. La revolution des neurons artificiels et de l'apprentissage profond*. Paris: Odile Jacob.

Lipson, H. 2019. "Robots on the Run." *Nature* 568: 174–5.

Mitchell, M. 2019. *Artificial Intelligence. A Guide for Thinking Humans*. New York: Farrar, Strauss and Giroux.

O'Neil, C. 2016. *Weapons of Math Destruction*. Washington: Crown Books.

Picard, R. 1997. *Affective Computing*. Cambridge, MA: MIT Press.

Rahwan, I., M. Cebrian, O. Obradovich, J. Bongard, J.-F. Bonnefon, C. Breazeal, J. Crandall, N. Christakis, I. Couzin, M. O. Jackson, N. Jennings, E. Kamar, I. Kloumann, H. Larochelle, D. Lazer, R. McElreath, A. Mislove, D. Parkes, A. Pentland, M. Roberts, A. Shariff, J. Tenenbaum, and M. Wellman 2019. "Machine Behavior." *Nature* 568: 477–86.

Rausch-Hindin, W. B. 1988. *A Guide to Commercial Artificial Intelligence. Fundamentals and Real-World Applications*. New York: Prentice Hall.

Ricoeur, P. 1965. *De l'interprétation. Essai sur Freud*. Paris: Seuil.

Solms, M. 1996, "Towards an Anatomy of the Unconscious." *Journal of Clinical Psychoanalysis* 5, no. 3: 331–67.

———. 2008. "Repression: A Neuropsychoanalytic Hypothesis." www.veoh.com/watch/v6319112tnjW7EJH

Sulloway, F. 1979. *Freud, Biologist of the Mind*. New York: Burnett.

Tauber, A. I. 2014. "Freud without Oedipus: The Cognitive Unconscious." *Philosophy, Psychiatry, & Psychology* 20, no. 3: 231–41.

Voosen, P. 2017. "The AI Detectives. As Neural Nets Push into Science, Researchers Probe Back." *Science* 357: 22–7.

Wilson, T. D. 2002. *Strangers to Ourselves. Discovering the Adaptive Unconscious.* Cambridge, MA: Harvard University Press.

Winston, P. H., and K. A. Prendergast. 1984. *The AI Business. The Commercial Uses of Artificial Intelligence*. Cambridge, MA: MIT Press.

Chapter 2

The unconscious and technology

2.1 Introduction[1]

Object relations theory (Klein, Bowlby, Winnicott, Fairbairn) has realized a radical revision of Freudian thought. However, object relationships are essentially understood as interpersonal relationships, not as relationships with inanimate objects or artifacts. Objects are other people, not artifacts. The role of technique and unconscious content crystallized in technical objects has never been adequately taken into consideration, either by psychoanalysis or by the philosophy of technology.

I argue here that technology plays an important role in the formation of the unconscious. To justify this thesis, I propose a reinterpretation of some concepts of Lacanian psychoanalysis through Latour's anthropology of science, i.e., the actor-network theory (ANT). Before proceeding with the reinterpretation of Lacanian concepts, I want to answer three questions:

> Why do I choose Latour's anthropology of science as a model?
> How do I interpret Latour?
> What is my critical approach to Latour?

My reading of Latour is, above all, philosophical.[2] What is the central point of Latour's philosophy? I propose a summary using two formulas: symmetrical ontology and the realism of resistance. These two aspects interest me most in Latour's work. The first aspect, which is emphasized by object-oriented ontology (Harman 2019), is expressed by the first proposition of *Irréductions*: "Nothing is, by itself, either reducible or irreducible to anything else" (Latour [1984] 2011, 158). This means that all entities are on exactly the same ontological footing. However, "entities" are not static substances; they are centers of force. "There are only trials of strength, of weakness. Or, more simply, there are only trials. This is my point of departure: a verb, 'to try'" (158).[3] Reality comprises a set of forces, trials, and resistance: "Whatever resists trials is real," and "The real is not one thing among others but rather gradients of resistance" (158). The "form" is the stabilization of forces, a dynamic equilibrium. "A form is the front line of a trial of strength that de-forms, trans-forms, in-forms or per-forms it. Of course, once a form is stable, it no longer appears to be a trial of strength" (224). Here, I see a profound analogy between Latour's ontology and Simondon's (2005) theory of individuation.

I do not wish to analyze all the philosophical implications of these propositions. My only objective is to emphasize that the "principle of irreducibility" leads Latour to an approach that equates humans and non-humans, giving priority to their dynamic interactions, the "translations" of one another. It is a "flat ontology" (Harman 2019, 54), i.e., an ontology that treats all objects in the same way, without a preexisting taxonomy, classification, or hierarchy.[4] Humans and non-humans are all "actors" in the same way within a network of associations where translation processes, i.e., exchanges of properties and skills, constantly occur (Latour 2005). Latour calls this kind of network *collectif* or "nature-culture."[5] It is the network that defines the actors and their relationships, not the opposite. This position avoids any kind of dualism: "Such has always been the strategy of the French sociologist and anthropologist: to show the matter of spirit and the spirit of the matter, the culture in nature and the nature in culture" (Romele 2019, 51).

There are also other reasons that led me to Latour. His anthropology and sociology avoid reducing scientific facts to simple social phenomena or forms of Kantian constructivism. His work also calls into question the classical epistemological approach to interpreting scientific facts, which he accomplishes by considering science "in action" (Latour and Woolgar 1979; Latour 1987). The result is a form of relativistic materialism, which arguably constitutes a very mature form of realism based on "a relational materiality" (Law 1999).

Nevertheless, my reading of Latour is also critical. To my mind, Latour does not adequately consider AI and the world of digital technology. I share the opinion according to which Latour "is for sure not a digital sociologist" (Romele 2019, 51) and that his views about the digital stand in contradiction to his main approach. To summarize, I do not believe that Latour fully develops his principle of "irreducibility." Had he done so, he would have tackled the AI problem from the beginning. We cope with a paradoxical situation: on the one hand, the Latourian concept of the *collectif* is very useful in explaining machine behavior and human–non-human contexts; on the other, in Latour, a complete and rigorous theory of AI is absent. AI is a field in which the difference between humans and non-humans, or rather, the difference between living and non-living beings, is radically challenged. Does Latour really reject the "modern compromise" (*l'Ancien Régime*, i.e., traditional metaphysics; see Latour 1991)? However symmetric ANT may be, it lends more focus to the actions of scientists and engineers. In Latour, humans always play the leading role in the constitution of non-humans: "the Pasteur network" creates the microbe; "the Joliot network" creates the nuclear chain reaction; the geologists in the Amazon create the forest, etc. (Latour [1999]

2007, 30–55). AI overturns this scheme: there are new forms of human–non-human hybrids in which it is the non-human part that exercises control of the situation and "creates the fact." AI is thus a real test for ANT.

In a nutshell, my criticism is that, in Latour, there is no real AI theory (see also Elliott 2018, 41). However, precisely because of its philosophical assumptions, his method can be applied to AI with advantageous results (even if with risks: see Jones 2017). In other words, Latour lacks cybernetics, i.e., a theory of intelligent machines in the sense of Foerster (1960) and Günther (1957). A critical discussion on cybernetics would yield considerable advantages to the anthropology of science and ANT.

This is my critical interpretation of Latour. (For a broader critique of Latour's ANT, see Sismondo 2014.) On this basis, I suggest a twofold extension of his approach, i.e., in the direction of AI and psychoanalysis. I shall show that it is possible to reinterpret some fundamental concepts of psychoanalysis by using Latour's anthropology of science. The remarkable feature emerging from this reinterpretation is that the human unconscious is essentially technical—it presupposes artifacts. I mean that the human unconscious is the effect of a technical mediation. My strategy is clear. I use Latour's anthropology of science as the mediation between psychanalysis and AI. From this point of view, I will show that the concept of an "algorithmic unconscious" is plausible and can be a useful tool for analyzing AI behavior.

2.2 The mirror stage: from technology to the unconscious

In this section, I propose a new interpretation of Lacan's mirror stage. First, I present a short description of the mirror stage according to Lacan's texts. Second, I develop a reinterpretation of Lacan's mirror stage in terms of Latour's anthropology of science.

2.2.1 Lacan's concept of the mirror stage

The mirror stage is the starting point of Lacan's psychoanalysis, i.e., in a historical and theoretical sense. It is a complex dynamic of looks, postures, movements, and sensations concerning children between six and 18 months of life (Lacan 1966a; Roudinesco 2009). The child does not speak and does not yet have complete control of his/her body: he/she can barely stand upright, cannot walk, and must be supported by an adult. The child is not autonomous. The infant's world comprises a kind of merging of itself and the maternal body, a body that provides satisfaction and pleasure. This is a fundamental anthropological fact; the human being is born with a body that is not immediately in control, is late in its development, and needs adults for a long time (see Bolk 1926, and the concept of "fetalization"). This pre-Oedipal phase is called by Lacan "imaginary order"; it is a peculiar realm of ideal completeness where there are no boundaries between outside and inside prior to differentiation and individuation.

In the essay "The mirror stage as formative of the function of the I" (1949), Lacan describes the infant's moment of recognition of itself in a mirror by referring to Freud's theory of narcissism. The mirror stage is a narcissistic process in which the child constructs a misrecognized image of self-unity. Whether the mirror stage is understood literally or metaphorically,

> the crucial point for Lacan is that the small infant is led to misrecognize and misperceive itself. According to Lacan, the mirror provides an illusory apprehension of self-unity that has not been objectively achieved. That is to say, the creation of an "ideal self"—the self as it would like to be, self-sufficient and unified—is an imaginary construct, wish-fulfillment, pure and simple.
>
> (Elliott 2015, 33)

When such a small child looks in the mirror, he or she first sees another child who reproduces his or her gestures and movements. The child and the other in the mirror are synchronized: if one raises an arm, the other does the same; if one twists his or her mouth, the other does the same, etc. The child then tries to get to know the other child but encounters unexpected resistance: the cold and homogeneous surface of the mirror, which is an impassable border. The child tries to get around the mirror, but this attempt fails. He or she cannot find the other child in the mirror. At this point, something happens that clearly changes the child's perception in that situation. The child's gaze turns in a different direction and sees the surrounding objects and the adult who supports him or her reflected in the mirror. Thus, the child realizes that he or she sees duplicates because the mirror produces a doubling of the real. Everything is reflected in the mirror, and the reflection doubles everything. Everything is doubled, except for one thing: a face, the face of the other child. Everything is doubled, except for that childish face. The child cannot find the mirrored child in the surrounding environment. For this reason, in this unique element, the child recognizes his or her face and realizes the first fundamental distinction between himself or herself and the rest of the world. Thanks to the mirror, the child perceives his or her body as a unity, and this fact gives the child satisfaction and pleasure. Even if the child cannot speak, he or she enjoys this moment.

In Lacan's view, the mirror stage reveals the tragedy of the human quest for identification. The child makes an imaginary identification with its reflected image; the mirror provides a gratifyingly coherent image of himself or herself as unified. Human beings can grasp their identity only by passing through the other, something external, an object: the mirror. The subject is originally alienated, divided, split. Furthermore, the image in the mirror is not only external to the subject; it is also false. The mirror distorts things. It gives us a reversed image of ourselves and things. This means that the child is separated from his or her identity; the identification of the ego (*Je*) must always pass through the other, the imago in the mirror (*moi*), the "ideal ego." Such is the paradox: in order to have an identity, the subject must alienate itself. The imago remains unreachable and risks turning into a paranoid delusion. In other words,

> the mirror stage is profoundly "imaginary" for Lacan because the consolingly unified image of selfhood generated is diametrically opposed to the multiplicity of drives and desires experienced by the child. In a word, then, the mirror *lies.* The reflecting image, because it is outside and other, leads the self to misrecognize itself.
>
> (Elliott 2015, 105)

The mirror stage, however, also teaches us another important thing: it is through the illusion and falsity of the image that selfhood takes shape. For Lacan, the mirror stage is not a phase that is overcome and abandoned. An imaginary and paranoid background is constantly present in subjectivity. This means that the imaginary illusion—as a place of indistinction between the self and the other, between narcissism and aggression—*is a necessity of the subject*. It is the beginning and end of all forms of identification. In other words, the human being has a relationship more with the imagination than with reality. This aspect has been underlined and developed, in particular, by Castoriadis (1975), who formulates a very important criticism of Lacan that we cannot analyze here. For Castoriadis, imagination is a creative, innovative force that shapes the subject, the social reality, and their autonomy.

With the mirror stage, Lacan offers us a materialistic theory of subjectivity. The ego (*moi*) is the image reflected in the mirror. Thanks to the mirror, the child uses his or her imagination to give himself or herself an identity. The child defines himself or herself through the imagination and tends to present himself or herself to others through the imaginative construction he or she created. The mirror stage is the initial moment that allows every intersubjective relationship through which the subject identifies itself. For Lacan, the psychic development of the human subject passes from the identification with the imago in the mirror to the imaginary identification with other persons. Following the mirror phase, the child passes through a series of identifications: first with the mother and then with the other members of the family, especially brothers or sisters. However, this process is a source of pain and instability: the "other" self is simultaneously the source of and a threat to the identification.

Lacan contrasts the symbolic order with the imaginary order. In the symbolic order, the libidinal relation between child and mother is ruptured by the intervention of social forces. The symbolic order includes social meanings, language, logic, differentiation, and individuation. The child knows the symbolic order—and, therefore, a different type of subjectivation—through the Oedipus complex.

In the Oedipus complex, the subject identifies with the father, who imposes the prohibition of the child's desire for the mother. By identifying with the father, the child no longer identifies with a mirrored image or with a similar one, such as the mother or a brother. Through the father, the child identifies with a language and culture. Like Lévi Strauss, Lacan thinks that the prohibition of incest is a necessary condition of society and culture. The father has a symbolic function, not an imaginary one. He ends the cycle of paranoid identifications. The identification with the father is symbolic, i.e., an identification with the symbol, with the language (Tarizzo 2003). However, the symbol also addresses the other: it is a request for recognition and the beginning of a new experience of desire. The patient is a person who has remained a prisoner of his or her primitive identifications. Healing consists of the passage from the un-symbolized imaginary to the symbolized imaginary, i.e., a limited imaginary that can be part of a family and a social contest (Rifflet-Lemaire 1970, 138).

2.2.2 *Latour's reinterpretation of the mirror stage*

Let us now try to reinterpret the primitive Lacanian scene in terms of Latour's concept of the *collectif*. We can begin by reading this passage from *The Pasteurization of France*:

> There are not only "social" relations, relations between human and human. Society is not made up just of humans, for everywhere microbes intervene and act. We are in the presence not just of an Eskimo and an anthropologist, a father and his child, a midwife and her client, a prostitute and her client, a pilgrim and his God, not forgetting Mohammed his prophet. In all these relations, these one-on-one confrontations, these duels, these contracts, other agents are present, acting, exchanging their contracts, imposing their aims, and redefining the social bond in a different way. Cholera is no respecter of Mecca, but it enters the intestine of the hadji; the gas bacillus has nothing against the woman in childbirth, but it requires that she die. In the midst of so-called "social" relations, they both form alliances that complicate those relations in a terrible way.
>
> (Latour 1986, 35)

We can say the same thing for psychoanalysis. Psychoanalysis is a technique, a central aspect in Freud and Lacan. This means two things: (a) artifacts mediate the relationship between the analyst and the patient (e.g., the setting – there are objects that have a regressive function, i.e., they allow the patient's regression); and (b) the technique acts in the formation of the unconscious—there is a technical mediation of the unconscious.

In the mirror stage scene, there are three *actors*: (a) the child, (b) the mirror, and (c) the objects (human and non-human) that surround the child and are reflected in the mirror. These actors are all on the same level: they are neither reducible nor irreducible to each other (*Irreductions*, 1.1.1). The actors define each other, ascribing to each other strategies and wills. The child is reflected in the mirror and, with this gesture, creates a network, an association. "Microbes were not merely entities that Pasteur studied, but agents with whom Pasteur built an alliance. The alliance was ultimately very successful" (Sismondo 2014, 291). Properties and qualities circulate among the actors. The child is an *entelechy* (1.3.1). It is a force that wants to be stronger than others and, therefore, enrolls other forces (1.3.2), i.e., the mirror and the objects that surround it, among which is also the adult who supports the child. In this relationship (1.3.4), each actor acts and undergoes trials and resistance.

In this perspective, we can no longer say that the imago constitutes the identity of the child. It is the connection between humans and non-humans that constitutes this identity. This means that the imago is not a mere visual or auditory perception; it is a complex series of mediations between humans and non-humans. The Lacanian imago (*moi*) is the product of technology, i.e., the effect of a technical object: the mirror. Hence the unconscious is the effect of technical and material mediations.[6] Technology produces the imaginary ego that must be repressed by language. We are overtaken by what we manufacture. The mirror is not a simple tool that acts as a connection between the subject and the imago. The mirror is an actor like others.

The same thing can be said about Freud. In *Beyond the Pleasure Principle*, the child—Freud's nephew—learns to cope with the absence of his mother through the use of a spool, a technical object, an artifact, which he pulls and draws to himself. In this game, Freud captures the phenomenon of repetition compulsion. The child expresses its libidinal renunciation of the mother through the toy. The importance of technology for psychoanalysis is also confirmed by Freud's considerations on the *Wunderblock* (the "Mystic Writing Pad"). Moreover, the centrality of the relationship with the object in the child's psychic development has been underlined by Winnicott (2005) through the concept of "transitional objects." The study of the connections between AI and child psychology is a new and highly interesting field.

Let us analyze the *collectif* [child + mirror + surrounding objects] through the categories that Latour describes in the sixth chapter of *Pandora's Hope* (*L'espoir de Pandore*). In this chapter, Latour distinguishes four levels of technical mediation: translation, composition, articulation, and black box. I will reinterpret the mirror stage through these four categories.

Each actor has an action program—a set of actions, intentions, goals, or functions—that clashes with that of other actors in the network. When this happens, there are two possibilities: (a) the cancellation of one of the forces, or (b) the merging of the forces and the creation of a new action program.

The condition of (a) and (b) is what Latour calls "translation" (Latour [1999] 2007, 188), i.e., a process of mediation and transformation of action programs with "the creation of a bond that did not exist before and that modifies, with more or less intensity, the two original terms" (188; my translation). In the translation process, each actor maintains its qualities, its action program, and its objectives, but a connection is built and a transformation takes place. An essential phenomenon occurs in this process: the passage of qualities and capacities from one actor to another. The child is no longer only a child but a child-in-front-of-the-mirror who receives from the mirror certain qualities and abilities: first, the ability to recognize the duplicates of the surrounding things, including of himself or herself. The mirror is transformed by the child; it is no longer a simple object; it is the place where the child looks for and finds his/her identity and, therefore, enjoyment. It is also the place of a privileged relationship between the child and the adult who holds him/her.

The mirror is no longer the mirror-resting-on-the-table; it becomes the mirror-in-front-of-the-child and, therefore, the mirror-instrument-of-identification. As Latour claims, humans and non-humans have no fixed essences: in the *collectif*, every actor undergoes a transformation of its qualities and abilities. The association [child + mirror] is a human–non-human hybrid. This hybrid expands later, including other actors, or even other human–non-human hybrids, i.e., the surrounding objects. These latter hybrids play a very important role in Lacan's *collectif* because it is thanks to their presence and reflection in the mirror that the child can identify himself/herself with the mirrored image of the "other" child. Thus, to read the mirror stage in Latourian terms means to overcome a rigid subject/object dualism and to understand the complexities of the dynamics of forces and resistances in the *collectif*. The final result of the mirror stage is the identification with the ideal ego, but this identification is accomplished neither by the child nor by the mirror nor by the surrounding humans and non-humans, but by the associations of them all. "Action is not simply a property of humans, but a property of an association of actors" (Latour [1999] 2007, 192).

There are two forms of translation. The first is called "composition," and it is the process of mutual adaptation of the actors. Every actor has an "action program" in the network. The child is reflected in the mirror and is pleased to see himself or herself, while the mirror produces images, and the other human and non-human hybrids interact in several different ways (the adult holds the child up and talks to him or her, which influences the child's experience, but the child can also be distracted by other objects reflected in the mirror, such as a toy, etc.). The process of translation and mutual adaptation between action programs goes on until the child recognizes himself or herself in the mirrored image. However, this is a precarious equilibrium. New identifications take place.

The second form of translation is called "articulation." By "articulation," Latour means that the sense of the actions within the network depends on human–non-human relations. Non-humans can play an active role in these relations. The sense of a child's actions is created by the mirror. The child is held by the adult in front of the mirror and looks at it. The mirror produces the doubling that makes the child's experience of auto-recognition and identification possible. By reflecting an image that includes the other human–non-human hybrids surrounding the child, the mirror leads the child to believe that the only image that is not a duplicate is his or her image, his or her duplicate. This process can be described as follows:

> mirror > mirrored objects (human and non-human) > child's auto-recognition
>
> > child's identification > distinction between ideal ego/external world > *imago*

The relationship with objects precedes and determines the identification process. There is no sovereign subject that creates meaning. There is instead a technical object (the mirror, an artifact) that produces what Latour calls an *articulation*, a specific connection between humans and non-humans that produces new meanings and identifications. Understanding the meaning of an action does not mean investigating the mind of the person who performed it. It means carefully analyzing the processes of translation, composition, and association between humans and non-humans in a given situation.

The last step of our scheme is the fracture between the ideal ego and the external world. The child becomes paranoid: he or she tends to identify himself or herself with an abstract imago and to separate himself or herself from the rest of reality. This fracture completely covers and eliminates the mediation between humans and non-humans that we have just described. Everything is reduced to the ideal ego and the subject/object dualism. Now, I interpret the mirror stage outcome by using Latour's fourth category, *la mise en boîte noire*, the black box, "an operation that makes the joint production of actors and artifacts totally opaque" (Latour [1999] 2007, 192–3). This Latourian category is very important, and I will say more on it in the last section. My thesis is that the unconscious is essentially a *mise en boîte noire*, a black box. Latour inadvertently gives us the keys to a new phylogenesis of the unconscious and, therefore, the beginning of a new kind of psychoanalysis.

Latour's idea is simple: the technical artifact hides the set of practices that constitute it. The final results (e.g., the microbe in Pasteur) produce a sort of paradoxical feedback: they cover and hide the human–non-human interactions as well as the processes and dynamics that produced them. Thus, the last phase of the work hides all the paths that lead to it. "When a machine works effectively, when a state of affairs is established, we are only interested in the inputs and outputs, not in their internal complexity. This is how, paradoxically, science and technology know success. They become opaque and obscure" (Latour [1999] 2007, 329). When a scientific fact or artifact is established and "closed," the *collectif* disappears by being crystallized in its outcome: it is "reduced to a single point," says Latour.

In Lacan's mirror stage, there are two black boxes: (a) the first coincides with the imago itself, which hides the mirror and the rest of the surrounding world. The imago produces the auto-recognition and identification, which are abstractions from the technical and material conditions that constitute them. The child removes the mediation of non-humans such as the mirror; therefore, he or she distinguishes himself or herself from them. In other words, the image that constitutes the child's identification is also what blinds the child and renders him or her incapable of grasping the imaginary nature of his or her identification. This first black box is closed and reopened many times: the child goes through a range of imaginary identifications. (b) The second black box is much more stable and coincides with the transition from the imaginary to the symbolic and, therefore, with the Oedipus complex. The symbolic (the Name of the Father; more on this later) "closes" the mirror stage, making it a black box. In fact, according to Lacan, the symbolic interrupts the series of imaginary identifications. This interruption coincides with the *Spaltung*: repression. The symbolic removes the imaginary, making it a symbolized imaginary. Reduced to a black box, the imaginary can be limited or removed. In the patient, this process is absent or incomplete.

2.3 The Oedipus complex: from the unconscious to technology

The Oedipus complex is the fundamental structure of emotional and interpersonal relationships in which the human being is immersed. Freud (2005, [1905] 2011, [1917] 2012) has hypothesized that the Oedipus complex occurs when the child is three to five years old. This psycho-affective organization is based on attraction toward the parent of the other sex and jealousy and hostility toward the parent of the same sex. The core of this organization is the prohibition of incest.

In the case of the young boy, the child develops sexual knowledge of the penis and fantasizes about sexual union with the mother. This fantasy of sexual union is, however, dispelled from outside the child/mother dyad by the father. The boy comes to hate his father's superior control of the maternal body and fantasizes about his death. However, recognizing that he cannot compete with the phallic authority of the father, and facing the imagined threat of castration (the so-called "castration complex"), the boy renounces his primary erotic investment, repressing sexual desire for the mother permanently into the unconscious. In the case of the girl, Freud introduces an additional element: the penis envy. Female sexuality arises from a denial of male sexuality. The fundamental model of sexuality for Freud is masculine.

The contestation of this model is the central point of the feminist debate on Freud.

According to Freud, the Oedipus complex is a fundamental element in the development of the human personality, and if it is not overcome, it constitutes the basic nucleus of all psychopathologies. The entire original phantasmal world of humans is related to the Oedipus complex. The formation of the super-ego is also seen as resulting from the introjection of the paternal prohibition against sexual relations with parents, brothers, and family members in general. For Freud, the Oedipus complex is the central point of sexual development, the symbolic internalization of a lost, tabooed object of desire. The Oedipus complex also reveals another important aspect of Freudian psychoanalysis: self-constitution is always the consequence of the loss of the object, and above all, the primary object, the mother. It is, therefore, an emotional (pain of loss) and imaginary (substitute for the lost object) process. For Freud, "self-constitution arises as a consequence of loss. Selfhood is formed under the sign of the loss of the object, in an attempt to become like the lost love" (Elliott 2015, 21). This connection (self-constitution + loss) is also pivotal for Lacan.

From a Freudian perspective, the Oedipus complex is the doorway through which the subject enters society, culture, and adult life. However, Freud's teaching on culture and the social world, in general, remains contradictory. Even if it is true that culture represents the sublimation and overcoming of the deepest impulses, there are still unconscious desires that resist social patterns and conventions. The work of psychoanalysis, that is, the digging into the depths of cultures, never ends.

I will proceed as follows. I will first present a short description of Lacan's interpretation of the Oedipus complex. This will not be an exhaustive analysis but only an overview of Lacan's interpretation of this primordial "scene." Second, I will reinterpret Lacan's interpretation by following Latour's anthropology.

2.3.1 *Lacan's Oedipus complex*

Following Lévi-Strauss (1955, 1962), Lacan claims that the prohibition of incest constitutes a universal law that differentiates human civilization from the state of nature. Lacan uses the expression "Name-of-the-Father," which defines the acceptance of social law and marks the passage from a potentially psychotic pre-human condition, that of the mirror stage, to a real human condition. In contrast to a reductive focus on the immediate family situation, Lacan "contends that the father intrudes into the child/mother dyad in a symbolic capacity, as the representative of the wider cultural network and the social taboo on incest. Not only is the child severed from the imaginary fullness of the maternal body, it is now inserted into a structured world of symbolic meaning—a world that shapes all interactions between the self and others" (Elliott 2015, 106). In Lacan's view, the psychotic has not internalized the Name-of-the-Father. The originality of this interpretation compared to Freud's is that the Name-of-the-Father does not coincide with the actual father.

In Lacan's view, the mother and child live a symbiotic relationship that breaks with birth. Both have a kind of nostalgia for this original condition and want to recreate it. Weaning is the traumatic phase: the contiguity between the bodies, maintained by breastfeeding, is interrupted. Both the mother and child have a regressive desire: they want to return to the situation of original dependence. This is a desire for identification; the child identifies himself or herself with the mother (Lacan 1966a, 35–45).

The father disrupts this situation and prevents the regressive impulse. As I said above, what characterizes Lacan's interpretation is that the paternal prohibition is considered in symbolic terms. Lacan thinks of the Oedipus complex as the structure of the sign. In general, this allows him to apply the Oedipus complex to both males and females.

The father represents the social law of coexistence and, therefore, the language. Lacan (1966b, 1998) introduces the concept of the Name-of-the-Father (*nom du père*), which is based on French homophony between *nom* (name) and *non* (no, negation), in order to highlight the legislative and prohibitive role of the father. The Name-of-the-Father is the original repression; it diverts the immediate and original mother and child impulse. In doing so, this repression opens the space of the sign, of the appearance of language. Lacan follows Heidegger: language precedes individuals, we are "spoken" and determined by language, and we were born into a universe of words over which we have no power.

Let us clarify this thesis. What Lacan has in common with many other important interpreters of Freud "is the claim that Freud's most original and important innovations were obscured and compromised by his effort to embed psychoanalysis in biology and thereby to scientize his vision of the psyche" (Mitchell and Black 1995, 195). Lacan rereads Freud's core concepts (the unconscious, repression, infantile sexuality, and transference) in the light of modern linguistics and post-structuralist theory. His famous maxim "The unconscious is the discourse of the Other" means, essentially, that human passion is constituted by the reference to the desire of others: both internal otherness (the unconscious) and external otherness (language). Our deepest unconscious feelings and passions are always expressed, as it were, through the "relay" of other people. As a result, psychoanalysis "is a theory about the fabrication of the individual subject as refracted through language and thus the social world" (Elliott 2015, 101). Lacan's starting point is the perpetual fragmentation of the subject; this, for Lacan, is Freud's discovery. However, Freud and ego psychology have not been able to draw all the consequences of the disruptive power of the unconscious. This conviction also explains Lacan's original writing style: to rediscover the radical nature of Freud's discovery requires an unusual form of writing and communicating. Reading Lacan is a difficult endeavor.

The core of Lacanian psychoanalysis is the connection between the rereading of Freud, linguistics, and post-modernism. For Lacan, language is the fundamental medium in which desire is represented and through which the subject is constituted. The claim "The unconscious is structured like a language" means that we are born into a language and that our desire is constantly immersed in it. We are not the inventors of language. Language is not just a tool to describe the world. It dominates and determines us at all times. It is independent of the subject, but it reveals the truth of the subject.

This conception cannot be understood without mentioning another crucial aspect: the fracture between *signifier* and *signified*, which Lacan takes from Saussure. According to the Genevan linguist, language is a system of signs, and every sign is made up of a signifier and a signified. The signifier is the phonological element of the sign; it is the *image acoustique* linked to a signified (i.e., the immaterial meaning). In line with structuralist linguistics, Lacan argues that the relationship between signifiers and signifieds is arbitrary and based on convention. It is a social structure. Meaning is created through linguistic differences and through the play of signifiers. That is to say, for instance, the meaning of a signifier—"man"—is defined by difference, in this case, with the signifier "woman." The relationship between sign and object is always provisional and arbitrary, and its usage depends upon historical and cultural conventions. The linguistic structure is organized along two axes: *condensation*, which can be conceived of as metaphor (synchronic order), and *displacement*, which can be equated with metonymy (diachronic order). In the first case, the signifiers are superimposed, juxtaposed, and synthesized, while in the second, an exchange takes place, namely, the substitution of one signifier for another. Metaphor and metonymy are the two axes along which the Lacanian unconscious operates.

Reinterpreting Saussure, Lacan introduces the following formula, which he defines as an "algorithm":

$$\frac{S}{s}$$

In this formula, "S" indicates the signifier and "s" the signified. The formula affirms the primacy of the signifier over the signified, i.e., the primacy of the normative, mechanical, and material dimensions of the language. The signifier is a meaningless material element in a closed differential system, i.e., the structure. Thus, the signified (and thus the meaning, the subject, and the ego) is only a secondary effect of the combination and recombination of signifiers. There is never a full, absolute signified. The signified is something that is always "pushed back" by the succession of signifiers and shaped by the "symbolic chain." Whereas Saussure places the signified over the signifier,

Lacan inverts the formula, putting the signified under the signifier, "to which he ascribes primacy in the life of the psyche, subject and society" (Elliott 2015, 106).

Lacan's main innovation is to interpret the Saussurian distinction between the signifier and signified in terms of *repression*. At the same time, the unconscious is what guides the game of combination and recombination of signifiers (the unconscious that speaks) and what is repressed and censored by signifiers (the unconscious that does not speak). The meanings produced by the symbolic chain tell us about a more fundamental meaning: enjoyment (*jouissance*), that is, the unconscious that does not speak. The symbolic chain is repression because it "saves" the subject from the enjoyment and allows him or her—through the analysis—to build a new relationship with the force of enjoyment, i.e., with the drive. What distinguishes the Lacanian linguistic structure from the Chomskian generative grammar is the fact that for Lacan, it is the pure impulse, the unconscious force of enjoyment that speaks in the structure. The unconscious is manifested in the artifact (language).

In other terms, in the passage from the mirror stage to the Oedipus complex, the subject passes from a state of absolute imaginary freedom to a condition of absolute constraint. Desire is symbolized, that is, enclosed in a system of differences and combinations based on social norms. Again, for Lacan, the self is always alienated from its own history, is formed in and through otherness, and is inserted into a symbolic network. The Name-of-the-Father (the first essential signifier) disrupts the imaginary union between the child and the mother—the enjoyment—and imposes social law. The statement "The unconscious is structured like a language" means that social, linguistic processes and the inner depths of the psyche are intertwined. To be a member of society

> requires a minimum level of linguistic competence, in order to adopt the position of speaker or listener. This demands the acceptance of a subject position in terms of the social conditions of culture, sexual difference and ideology. [...] The symbolic field—that is, language—is the crucial means through which individuals are "subjected" to the outer world.
>
> (Elliott 2015, 33)

The ancestral repression takes place in every linguistic act. The signifier is like a strongbox; there is a bomb in this strongbox, and this bomb is the enjoyment. The psychoanalytic process intends to open the strongbox and disarm the bomb. It is only through the interpretation of the symbolic chain and the game of signifiers that the patient can recognize his or her desire and enter a relationship with enjoyment in a good way; thus, the symptom becomes *sinthome* (Di Ciaccia 2013).

Let us summarize the results of our analysis. Language is the result of a deviation. Lacan reinterprets the Oedipus complex from a linguistic point of view. It is not the language that produces the deviation of the desire, but the reverse: the deviation of the primitive desire for identification is the cause of the emergence of the sign and language, that is, of what Lacan calls the "signifying chain." What do we see here? An instrument, a technology, namely, language, derives from an unconscious dynamic. The unconscious produces a technique and, therefore, also new cognitive abilities. The mirror and the language show that the unconscious is externalized in artifacts that strengthen or simulate our activities, emotions, etc. The Lacanian interpretation of the Oedipus complex is much less "mythical" and sex-focused than the Freudian one. Indeed, it helps to "demythologize" Freud's point of view and provides a critique of it.

2.3.2 Latour's Oedipus complex

Here, our reconstruction of Lacan's thought ends and its reinterpretation using Latour's terminology begins. As I said before, the child wants to reconstruct the symbiosis with the mother (and vice versa), but his or her desire is blocked by the father. Consequently, a deviation of desire takes place. In Latour's terminology, this deviation of the child's action program opens the door to intervention from another actor, which is the language, i.e., a technology, an artifact (Ong 1982).

For Latour, language is not only a set of symbols connected by deterministic rules; it is an actor similar to all other human and non-human actors (Latour [1999] 2007, 150). Language does not imply any abstraction from the world; rather, it is rooted in the world and has meanings, thanks to the connections among actors. This view is simultaneously very close to and very far from that of Lacan. It is very close because Lacan also believes that language surpasses humans and envelops them. It is very far because Latour does not rigidly interpret language as a game of differences based on inflexible rules. For Latour, Lacan is still a victim of the "modern compromise."

In light of the above, we can describe the Oedipus complex as a *collectif* composed of four actors [child + father + mother + language]. The action programs of the child and language connect each other. Therefore, a process of translation and mutual adaptation begins. From a Lacanian point of view, the purpose of the child is reunion with the mother, while the purpose of language is the signifying chain, i.e., the selection, combination, and recombination of signifiers. In the translation process, an exchange of qualities and abilities takes place in both directions. Both actors transform themselves. This means that if the child becomes the language, *the language becomes the child.*

Let us analyze both sides of the last statement. The child is no longer *infans*: he or she enters into a connection with the language, then he or she comes to know a new type of desire. Thanks to language, the child can dominate his or her own narcissistic impulses and live with other human beings in a community. Thanks to language, the child abandons his or her earlier narcissism. As a result, the child experiences finiteness. The situation can be

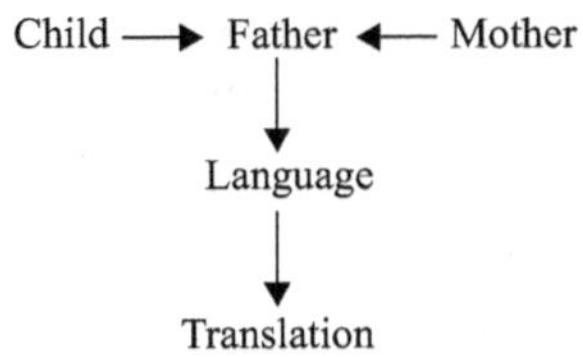

Figure 2.1 A Latourian reinterpretation of Lacan's theory.
The Oedipus complex can be described as a *collectif* composed of four actors: child, father, mother, and language.

described by Figure 2.1. The intervention of the non-human artifact resolves the contrast between humans. Language then becomes a black box in which these tensions are locked up and removed.

However, this reinterpretation still lacks an essential point. Following Latour, we must also go in the other direction: if the child becomes the language, *the language becomes the child.* A technical object—the language—acquires some of the child's characteristics. A translation takes place. What does it mean?

This is the crucial point of our reinterpretation. An artifact can assimilate, i.e., simulate, a human strategy. While the child assimilates the repressive mechanism of a significant chain, the language assimilates the child's search for identification; *therefore, it tries to identify with humans—it imitates humans*. The identification—following Lacan—becomes symbolic. However, this does not mean that language seeks identification for itself. The identification process always remains unidirectional: from humans to language. However, language does not play a passive role in this process. If we follow Latour's symmetrical anthropology, then we must admit the occurrence of an exchange of properties in both directions.

An objector could easily reply that we are attributing human qualities to non-humans indiscriminately and that this has contradictory consequences. Obviously, I am not saying that language acquires human qualities in a magical or animistic way. The *collectif* is not a metaphysical entity but a dynamic theoretical model that explains the associations of human and non-human actors. When Latour speaks of humans and non-humans, he does not aim to designate subjects and objects. The concept of the *collectif* arises from a profound critique of traditional metaphysics and its classic pairs of opposites: subject/object and fact/value. While the subject and the object are "closed" and opposing entities, the human and the non-human are instead "open" entities that are in constant interaction and mutually defining each other. Non-human entities are not opposed to humans; they make humans what they are, and vice versa. "Being human" entails a multifaceted relationship with non-human entities and organisms. Can a virus speak? Through science, yes. Latour says that the scientist is the "spokesman" of the virus, in the sense that the virus receives a social and intellectual role in the human world through scientific work. This is not a metaphor: the scientist gives voice to the facts through his or her research and the texts (papers, books, reports, documents, etc.) he or she produces. The fact does not speak, but the scientist gives it a voice. This is what happens in the laboratory. Moreover, the fact gives the scientist certain qualities by allowing him or her to discover new things and "directing" his or her research. For example, weapons are not mute objects at all, quite the contrary. Although indifferent to human passions, weapons arouse and convey them—or annihilate them. While lacking a will, weapons can shape and bend wills. The *collectif* is a crisscrossing and constant mutual translation of human and non-human capacities.

Latour (2004, chapter 4) distinguishes four phases of a *collectif*: perplexity, consultation, hierarchy, and institution (see Figure 2.2).[7] As noted above, the *collectif* is an association process. The moment an entity (human or non-human) wants to join it, that is, enter the *collectif*, it must pass a test to see if its "candidacy" is acceptable by the other actors of the *collectif* (this is the "perplexity" phase). If accepted, the "candidate" is challenged in order to understand whether and how it is compatible with the other actors that compose the *collectif* (this is the "consultation" phase). If admitted, the "candidate" receives a classification (a place in a hierarchy) and an essence; it becomes a stable part of the institution-*collectif*. Perplexity–consultation–hierarchy–institution compose a life cycle of the *collectif*. It is not a metaphysical abstraction. A laboratory works exactly like this; the scientific fact is the product of a series of activities and resistances of humans and non-humans that interact. The same can be said for the constitution of the law (Latour 2013). The *collectif* model helps us understand the complex dynamics of human-non human systems. Each actor acts (from the outside or the inside of the *collectif*) and causes a change in the *collectif* and in the other actors. The *collectif* is a set of forces in search of equilibrium.

Here, there is no extravagant projection of human qualities on non-humans; rather, there is only the overcoming of a naïve metaphysics that presupposes that objects are indifferent to humans and immutable. Latour

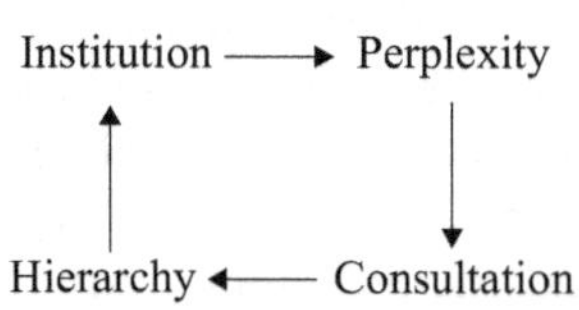

Figure 2.2 The four stages of the *collectif* model.
In Latour, the concept of *collectif* arises from a profound critique of traditional metaphysics and its classic pairs of opposites: subject/object and fact/value.

politicizes traditional metaphysical categories: subject and object are created notions, which represent certain forms of agreement between humans and non-humans. Subject and object "are the names given to forms of representative assemblies so that they can never meet together in the same chamber" (Latour 2004, 111; my translation). These are not metaphors. Furthermore, the accusation of relativism is inconsistent. External reality is the reality external to the *collectif*, which the *collectif* accepts in itself or not. In other words, there are no things but rather systems or networks of things, which form ever larger systems whose borders are mobile. Reality is no longer an inert and inaccessible fact. It is what resists its assimilation by the *collectif*, either from inside or outside the *collectif* itself.

The therapeutic relationship and the development of the psyche can be described as a *collectif*. Patient and analyst are actors in a *collectif* who negotiate the admission of other members (human and non-human). The stability of imaginary identification is broken by the entry of language and social law into the child's *collectif*. The patient is the one who cannot admit language into his or her *collectif*. The patient cannot fully express his or her desire and story. For this repression, he or she suffers. The psychoanalyst is the one who must carry out and conclude negotiations successfully in order to allow language and memories to enter the patient's *collectif*. The patient heals when he or she "speaks" about his or her memories and story, that is, when he or she gives voice to his or her desire and the prescriptions of the social law.

The same thing happens in the Oedipus complex. According to Lacan, the child is dominated by a fundamental anguish: the fragmentation of the body and the separation from the mother's body. The child responds to these anxieties with a cycle of imaginary identifications. The Name-of-the-Father stops this cycle. Thanks to the Name-of-the-Father, the *collectif* changes. Language gives the child the social dimension and makes him or her an adult, while the child entrusts his or her anxieties to language, abandoning them by "unloading" them into language. For this reason, the unconscious continues to speak in language. Language "becomes human" because it becomes the locus of unconscious drives and emotions. Language is not a simple set of neutral symbols, a mute set of signs. It is an actor like others in the same *collectif*. It arises from paternal castration and social law; however, thanks to interactions with humans, it acquires human characteristics and takes part in the social game. Stating that language can assimilate human characteristics is tantamount to deleting the distinction between *langue* and *parole*, i.e., the structure and act introduced by Saussure (1949). This distinction does not exist at all. We have to think of the relationship between *langue* and *parole* as a translation process that goes in both directions. *Langue* is translated into *parole* and *parole* into *langue*. Language is a perennial negotiation between these two levels. These actors shape each other.

2.4 Conclusions

In Lacan, an unconscious dynamic of identification (the stage of the mirror and its endpoint, the Oedipus complex) leads to an artifact, i.e., the language. This artifact is conceived of as a machine, a chain of signifiers that combine with each other. As a result, language is the first AI—and perhaps the only one. This technology acquires human abilities and qualities and is able to reproduce them. We have thus isolated the first psychoanalytic sense of AI. The unconscious is not just the effect of technical mediation. It also produces technological effects. "The unconscious, as Freud tirelessly reminds his readers, transcends boundaries and influences the furthest reaches of our cultural and political lives" (Elliott 2015, 2).

Notes

1 Some parts of this chapter are a new development of Possati (2020).
2 By saying this, I do not classify Latour. I am simply saying that *my point of view* on his work is philosophical. Latour remains an unclassifiable thinker.
3 It is for this reason that I see a "priority of the controversies" in Latour's (2005, 25) sociological methodology.
4 This does not mean that I support the object-oriented ontology (OOO). I share only two aspects of OOO—at least in Harman's version—namely, the flat ontology and the principle of resistance (Harman 2019, 256–8). For the rest, I believe that OOO does not grasp the crucial point of Latour's philosophy, that is, its complexity. The rationality expressed in Latour's anthropology of science is one that starts from disorder, chaos, and uncertainty. The order is always a later result—rationality emerges from disorder. Harman grasps neither the centrality of the translation category nor the complexity of mediation. In Latour, mediation means combat, negotiation, and controversy. Latour's rationality is based on the Go model, not the chess model. The former is based on a few rules and visual and "cartographic" intelligence; the latter is based on many rules and a rigid space, the chessboard. In contrast, the *goban* is larger and the play more fluid.
5 The *collectif* is not a society: this is an essential distinction for Latour. The concept of society is still dependent on the traditional metaphysics based on the rigid dualisms subject/object and facts/values (see Latour 2004, 361–2).
6 An objector could reply that the mirror is only an accidental element in Lacan. True identification occurs in the child's encounter with the mother. However, the mother cannot be seen by the child as another human subject because, following Lacan, the child still lacks the linguistic relationship. The mother is considered an object among many others: a material object that helps the child produce its identifying image.
7 These phases are perfectly consistent with the categories described above: translation, composition, articulation, and black box.

References

Bolk, L. 1926. *Das Problem der Menschwerdung*. Jena: Fischer.
Castoriadis, C. 1975. *L'institution imaginaire de la société*. Paris: Seuil.

Di Ciaccia, A. 2013. “Il godimento in Lacan.” In *La Psicoanalisi. Studi internazionali del campo freudiano*. www.lapsicoanalisi.it/psicoanalisi/index.php/per-voi/rubrica-di-antonio-di-ciaccia/132-il-godimento-in-lacan.html
Elliott, A. 2015. *Psychoanalytic Theory*. London/New York: Palgrave Macmillan.
———. 2018. *The Culture of AI: Everyday Life and the Digital Revolution*. London/New York: Routledge.
Foerster, H. von. 1960. “On Self-Organizing Systems and Their Environment.” In *Self-Organizing Systems*, edited by M. C. Yovits and S. Cameron, 31–50. New York: Pergamon Press.
Freud, S. 2005. *The Unconscious*. London: Penguin.
———. (1905) 2011. *Three Essays on the Theory of Sexuality*. Reprint, Eastford: Martino Fine Books.
———. (1917) 2012. *A General Introduction to Psychoanalysis*. Reprint, Hertfordshire: Wordsworth.
Günther, G. 1957. *Das Bewusstsein der Maschinen: eine Metaphysik der Kybernetik*. Krefeld, Baden-Baden: Agis-Verlag.
Harman, G. 2019. *Object-Oriented Ontology*. London: Penguin.
Jones, R. 2017. “What Makes a Robot ‘Social’?” *Social Studies of Science* 47, no. 4: 556–79.
Lacan, J. 1966a. *Ecrits I*. Paris: Seuil.
———. 1966b. *Ecrits II*. Paris: Seuil.
———. 1998. *Le séminaire: les transformations de l’inconscient*. Paris: Seuil.
Latour, B. (1987) 1989. *La science en action. Introduction à la sociologie des sciences*. Paris: La Découverte.
———. 1991. *Nous n’avons jamais été modernes*. Paris: La Découverte.
———. (1984; Engl. trans. 1986) 2011. *Pasteur: guerre et paix des microbes, suivi de Irréductions*. Paris: La Découverte.
———. 2004. *Politiques de la nature*. Paris: La Découverte.
———. 2005. *Reassembling the Social*. Oxford: Oxford University Press.
———. (1999) 2007. *L’espoir de Pandore*. Paris: La Découverte.
———. 2013. *Making the Law*. Cambridge, MA: Polity Press.
Latour, B., and S. Woolgar. 1979. *Laboratory Life. The Social Construction of Scientific Facts*. Los Angeles: Sage.
Law, J. 1999. “After ANT: Complexity, Naming and Topology.” In *Actor-Network Theory and After*, edited by J. Law and J. Hassard, 1–14. Oxford: Blackwell.
Lévi-Strauss, C. 1955. *Tristes Tropiques*. Paris: Plon.
———. 1962. *La pensée sauvage*. Paris: Plon.
Mitchell, S., and M. Black. 1995. *Freud and Beyond. A History of Modern Psychoanalytic Thought*. New York: Basic Books.
Ong, W. 1982. *Orality and Literacy*. London/New York: Routledge.
Possati, L. M. 2020. “Algorithmic Unconscious: Why Psychoanalysis Helps in Understanding AI.” *Palgrave Communications* 6: 70.
Rifflet-Lemaire, A. 1970. *Jacques Lacan*. Brussels: Dessart.
Romele, A. 2019. *Digital Hermeneutics. Philosophical Investigations in New Media and Technologies*. London/New York: Routledge.
Roudinesco, E. 2009. *L’histoire de la psychoanalyse en France—Jacques Lacan*. Paris: Hachette.
Saussure, F. 1949. *Cours de linguistique general*. Paris: Payot.

Simondon, G. 2005. *L'individuation à la lumière des notions de forme et d'information*. Paris: Millon.
Sismondo, S. 2014. "Actor-Network Theory: Critical Considerations." In *Philosophy of Technology. The Technological Condition*, edited by R. C. Scharf and V. Dusek, 289–96. Oxford: Wiley.
Tarizzo, D. 2003. *Introduzione a Lacan*. Rome/Bari: Laterza.
Winnicott, D. 2005. *Playing and Reality*. London: Routledge.

Chapter 3

The difficulty of being AI

3.1 Introduction

In this chapter, I apply the results of the previous sections to the study of AI. One of the central theses of Freudian and Lacanian psychoanalysis is that the human psyche is the result of a series of identifications. The need for identification is a crucial element in human development. As we have seen, the mirror stage and Oedipus complex are forms of identification mediated by technology. The central question of this chapter concerns the type of identification at the origin of AI. My hypothesis is that the unconscious root of AI is a projective identification process. Projective identification is a concept that is currently much discussed in psychoanalysis. In outlining its structure, I mainly follow Bion's (1962) and Ogden's (1982) interpretation. Bion (1962) conceives of projective identification not only as a defense mechanism but also as a form of unconscious communication and tool of clinical analysis. I extend the use of this notion to the study of artifacts and AI. My claim is that projective identification is a useful tool to analyze and better understand the behavior of AI systems. I think that the great utility of projective identification as a criterion lies in its potential use in distinguishing between the often conflated real possibilities of AI and its imaginary aspects. Our AI fantasies often go beyond AI reality. We ascribe more to AI than it really is, and this imaginative process is the fruit of projective identification.

3.2 Projective identification in psychoanalysis

The concept of identification is central in psychoanalysis. Freud ([1917] 2012a) stresses the connection between identification and object investment. Identification is the most primitive and original form of emotional bond. Freud essentially distinguishes between two forms of identification: (a) primary identification, i.e., an original form of emotional bond with an object, e.g., a mother or father; and (b) secondary identification, i.e., the substitute for an abandoned object bond, which is constituted by the ego through introjection. The first form of identification takes place in the presence of the object, while the second occurs in its absence. In Freud (2012b), primary identification is described as an original dynamic of the psyche that precedes the Oedipus complex. The human being identifies with an object in the sense that he or she acquires a quality or set of qualities of that object, e.g., the mother or father. Lacan follows the same line: identification responds to a fundamental need of the child, i.e., for unity with itself and with the mother's body.

There is no single theory of identification in Freud: "it becomes abundantly clear that he [Freud] often uses the same terms to refer to what are basically quite different concepts"; this confusion "becomes still further confounded in the later developments and modification of Freud's theories by contemporary writers, who typically apply the same terminology in still other ways and introduce new names for concepts and processes discussed by Freud himself" (Bronfenbrenner 1960, 15). However, through all the transformations of the concept, there is one constant. Identification is always based on "an emotional tie with an object," typically parents. Parents are the fundamental object models. Identification is carried out with objects: a person or a person's trait (partial objects). Thus, Freud distinguishes between *defensive identification*, which is identification with the aggressor in order not to feel threatened by another person's power, and *anaclitic identification*, which is motivated by the desire to be similar to a loved one. These two types of identification are found in the Oedipus complex and the emergence of the conscience.

The Oedipus complex is a form of identification. Moreover, it is preceded and followed by other identifications. As Freud explains in "Mourning and Melancholia" (1917), in the first phase of identification, the ego and objects are merged into a single undifferentiated pattern. There is no distinction between the ego and the objects. This is anaclitic identification. The child identifies with both parents and introjects parts of them. In the second phase, libido is added to the object relationship. The ego identifies with the *loved* object, the mother (in the case of a boy). This means that the child tends to assimilate the qualities of the mother. This is an anaclitic and pre-Oedipal identification: the child wants to incorporate the mother or part of the mother into himself. In the third stage, libido has to deal with the injury caused by the father's prohibition: the identification assumes a hostile form and becomes the wish to replace the father.[1] The child identifies with the aggressor, i.e., the father. The ego introjects this emotional dynamic, which becomes part of itself (ego-ideal and super-ego), and the free libido is withdrawn into the ego. The ego-ideal and super-ego are the substitutes for a failed identification—a form of intellectualization.

This is the same three-phase sequence found in another important text, *Group Psychology and the Analysis of the Ego* (1921), where "for the first time the process [of identification] is represented explicitly as a mechanism for resolution of the Oedipus complex" (Bronfenbrenner 1960, 17). Through emotional connection, identification allows the transfer of the parents' qualities and abilities to the child's ego. Two aspects characterize Freud's theory: identification is always introjective, and it concerns the relationship between humans.

Even more interesting is the concept of projective identification proposed by Klein (1946) and Bion (1959, 1962, 1967). Klein and Bion contest many of Freud's and Lacan's ideas. While Freudian identification is introjective (the child applies the quality of the other person to himself or herself), the identification of Klein and Bion is projective, i.e., it occurs through the projection of the child's emotional content onto the mother (Spillius and O'Shaughnessy 2012). My hypothesis is that the desire for identification—in Klein and Bion's sense of projective identification—constitutes the deepest level of the algorithmic unconscious. AI is based on a new type of projective identification, i.e., one directed to artifacts.

For Bion, projective identification is a form of dialogue, an interaction between the child and mother, in which an exchange, a relationship of transformation, and mutual acceptance take place. Bion considers projective identification as the most important form of interaction between the patient and therapist in individual therapy and among groups of all types.

The following points regarding Bion's position are of great interest:

Projective identification is, first of all, an imaginary and emotional process: what is projected are *images connected to emotions*.

It is not only a fantasy but a manipulation of one person by another and, thus, interpersonal interaction.

The fundamental moment in psychic development is the confrontation with reality, which is a source of frustration. Reality presents itself to the child in a fragmented way. The infant is faced with an extremely complicated, confusing, and frightening barrage of stimuli. In making efforts toward organization, "the infant discovers the value of keeping dangerous, painful, frightening experiences separate from comforting, soothing, calming ones" (Ogden 1982, 21). Bion (1961, 1962) calls the baby's painful experiences (fear, hate, envy, aggression, etc.) "beta elements." The only way the infant can manage the beta elements is to project them outward, i.e., toward the mother. The child looks for a "container" in which to expel painful experiences. This kind of "splitting" is established as a basic part of the early psychological modes of organization and defense.[2]

However, how does this mechanism differ from simple Freudian projection? Bion claims that the child transmits its unbearable beta elements to the mother, who has the task of welcoming and "purifying" them from worrying aspects. Through her empathy, the mother must teach the child how to experience reality. Thus, the mother imbues the child's beta elements with meaning before returning them to the child. In doing so, she transforms the beta elements into what Bion calls "alpha elements," i.e., thoughts, representations, and content that can be used both in dreams and waking. The mother plays an active role: she must transmit to the child not only meaning but also the ability to act in the future on the same distressing content and "purify" it—to accept it. Therefore, what characterizes projective identification is not only the imaginary projection of the child but, above all, the acceptance and response of the mother. The child needs to be accepted and loved. A mother's acceptance and empathy are what Bion calls the ability of *rêverie*, which is one of the factors of the mother's "alpha function." The mother loves her baby through *rêverie. Rêverie* is the mother's ability to welcome complexity and clarify it so as to impose order on disorder. If the mother does not fulfill her task to the fullest extent, the child will become neurotic.

Projective identification, however, does not exclusively concern the mother–child relationship. Thanks to Bion's contribution, projective identification has assumed great therapeutic and heuristic value in psychoanalysis. This unconscious mechanism can be recognized and analyzed in processes such as transference and countertransference. "The analyst feels that he is being manipulated so as to be playing a part, no matter how difficult to recognize, in somebody else's phantasy" (Bion 1959, 149).

Bion uses the term "countertransference" to describe the analyst's emotional response to the patient's projective identification. This emotional response is a source of information about the patient and, therefore, can be used as an interpretative tool. For instance, Bion (1961) describes how, during a session with a patient, he began perceiving a growing fear of being attacked. He interpreted this experience by saying that the patient was pushing into his (Bion's) internal world the fear of being killed. He told the patient this interpretation, and the tension eased. Nevertheless, the patient was still clenching his fists. Bion then interpreted that the patient was again experiencing extreme fear, so much so that Bion feared that the patient was about to launch a murderous attack. During therapy, the analyst perceives his own emotions as a response to something external. These emotions are the consequences of a projective identification process enacted by the patient. The analyst's task must be to differentiate his or her own emotions from those of the patient and understand what originates in him or her and what does not. Errors in technique can reflect a failure on the part of the therapist to contain the patient's projective identification.

Let us clarify what we have said so far. Projective identification is the main way in which the child manages his or her emotional life. This confirms a profound intuition of Lacan: the human being is shaped by its imagination—in this sense, it is essentially paranoid. If projective identification encounters difficulties, the child's whole development is affected, with the risk of developing neurosis. According to Ogden (1982),[3] projective identification is an unconscious process. It is also (1) a type of defense against distressing content, (2) a way of communicating, (3) a primitive form of object relationship, and (4) a path toward psychological transformation, i.e., the maturity of the individual. Ogden distinguishes three phases of projective identification:

> In a schematic way, one can think of projective identification as a process involving the following sequence of events. First, there is the unconscious fantasy of projecting a part of oneself into another person and of that part taking over the person from within. Then, there is a pressure exerted through the interpersonal interaction such that the recipient of the projection experiences pressure to think, feel, and behave in a manner congruent with the projection. Finally, after being "psychologically processed" by the recipient, the projected feelings are reinternalized by the projector.
>
> (Ogden 1982, 12)

Let us better analyze these three phases.

First, let us consider the fantasy of projecting part of ourselves (emotions, feelings, qualities, and parts of the body) onto another person. The imaginary is connected to emotion and anxiety, i.e.,

> wishes to rid oneself of a part of the self (including one's internal objects), either because that part threatens to destroy the self from within or because one feels that the part is in danger of attack by other aspects of the self and must be safeguarded by being held inside a protective person. [4]
>
> (Ogden 1982, 12)

Projective identification can also be considered a form of repression by which the ego is relieved of the burden of having to bear certain emotional content that weighs on it. It is important to go to the root of the act: "This type of fantasy is based on a primitive idea that feelings and ideas are concrete objects with lives of their own" (Ogden 1982, 13). The fantasy of putting part of oneself into another person and controlling that person from within "reflects a central aspect of projective identification: the projector is operating at least in part at a developmental level wherein there is profound blurring of boundaries between self and object representations" (Ogden 1982, 13).

In the second phase, the fantasy of splitting off a part of the self becomes an act and, consequently, a real event. The projector pressures the recipient into making this split feel and occur according to his or her fantasy. "This is not an imaginary pressure, but rather, real pressure exerted by means of a multitude of interactions between the projector and the recipient. *Projective identification does not exist where there is no interaction between projector and recipient*" (Ogden 1982, 14).[5] The projector awaits confirmation that the harmful parts have been sent to the recipient. When the projector receives this confirmation, he or she experiences relief. From this point of view, projective identification is a form of paranoid manipulation. This especially happens in small groups, such as families. One member of a family can manipulate reality in an effort to coerce another member into "verifying" a projection, thus undermining reality testing. "Reality that is not useful in confirming a projection is treated as if it did not exist" (Ogden 1982, 16). However, what if the recipient does not answer the projector? A very interesting case—extensively studied by Ogden (1976)—is the projective identification of the mother on the baby "and the ever-present threat that if the infant fails to comply, he would cease to exist for the mother" (Ogden 1982, 16). The non-accepted baby would most likely develop a neurosis.

The third phase is the reinternalization of the projected content. This is the most delicate phase in the process. "In this phase the recipient experiences himself in part as he is pictured in the projective fantasy" (Ogden 1982, 17). The recipient experiences feelings that are his or her own; they are similar but not identical to those of the projector who influences him or her. Now,

> if the recipient can deal with the feelings projected into him in a way that differs from the projector's method, a new set of feelings is generated. This can be viewed as a processed version of the original projected feelings and might involve the sense that the projected feelings, thoughts, and representations can be lived with, without damaging other aspects of the self or of one's valued external or internal objects.
>
> (Ogden 1982, 17)

Therefore, *only if* the recipient is able to "digest" the feelings that he or she feels and that have been induced by the projector does something change. The "good" recipient shows that one can live with that anguish or even manage to give it a positive value. In this case, however, the projector stops the projective identification process and is able to accept his or her feelings. This is the process of "reinternalization." "To the extent that the projection is successfully processed and reinternalized, genuine psychological growth has occurred" (Ogden 1982, 18).

Projective identification is not only a metapsychological concept; it is also a real phenomenon with a circular structure (Figure 3.1).

In order to clarify this concept even further, the following example from Ogden (1982, 18–20) is illuminating as it clearly shows the three stages of identification. It is a long but important quotation:

> Mr. K. had been a patient in analysis for about a year, and the treatment seemed to both patient and analyst to have bogged down. The patient repetitively questioned whether he was "getting anything out of it" and stated, "Maybe it's a waste of time—it seems pointless," and so forth. He had always paid his bills grudgingly but had begun to pay them progressively later and later [...]. Gradually, the analyst found himself having

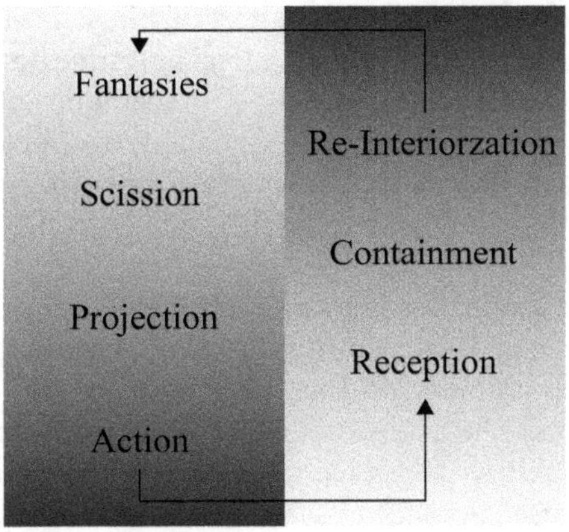

Figure 3.1 The main phases of projective identification.
Projective identification is the way in which the child manages his or her emotional life. This confirms a profound intuition of Lacan: the human being is shaped by its imagination—in this sense, it is essentially paranoid. Projective identification is a form of communication. There are different kinds of projective identification that need to be recognized in the analytic situation. The major distinction is between projective identification as a means of communication and projective identification as a way of expelling unbearable elements of the personality. The former is more positive. The second is much more characteristic of the severely disturbed, psychotic patient (Frosh 2012, 70).

difficulty ending the sessions on time because of an intense feeling of guilt that he was not giving the patient "his money's worth." [...] The analyst gradually began to understand his trouble in maintaining the ground rules of the analysis: he had been feeling greedy for expecting to be paid for his "worthless" work and was defending himself [...]. With this understanding of the feelings that were being engendered in him by the patient, the analyst was able to take a fresh look at the patient's material. Mr. K.'s father had deserted him and his mother when the patient was 15 months old. Without ever explicitly saying so, his mother had blamed the patient for this. The unspoken shared feeling was that the patient's greediness for the mother's time, energy, and affection had resulted in the father's desertion. The patient developed an intense need to disown and deny feelings of greed. He could not tell the analyst that he wished to meet more frequently because he experienced this wish as greediness that would result in abandonment by the (transference) father and attack by the (transference) mother that he saw in the analyst. Instead, the patient insisted that the analysis and the analyst were totally undesirable and worthless. The interaction had subtly engendered in the analyst an intense feeling of greed [...]. For the analyst, the first step in integrating the feeling of greediness was perceiving himself experiencing guilt

> and defending himself against his feelings of greed. [...] As the analyst became aware of, and was able to live with, this aspect of himself and of his patient, he became better able to handle the financial and time boundaries of the therapy. [...] After some time, the analyst's acceptance of his hungry, greedy, devouring feelings, together with his ability to integrate those feelings with other feelings of healthy self-interest and self-worth, was made available for internalization by the patient.

Why does the reception transform projective emotional content? Why does the receiving "save" one from anguish? By accepting the content within, the recipient shows the projector the possibility of accepting such content and living with the distressing emotions. The essence of what is therapeutic for the patient "lies in the therapist's ability to receive the patient's projections, utilize facets of his own more mature personality system to process the projection, and then make the digested projection available for reinternalization through the therapeutic interaction" (Ogden 1982, 20). The acceptance changes the value attributed to those distressing emotions and generates new emotions and new content (imaginings, representations, etc.). The recipient achieves this precisely by containing the distressing content. The projector feels welcomed. He or she sees that it is possible to accept the anxieties and live with them in a positive way. Thus, the projector gains maturity because he or she learns to do one fundamental thing: to accept and love himself or herself.

As we can see, the concept of projective identification is a useful tool in therapy. It provides a guide to the therapist to organize and render meaningful the data collected in the analysis. In many cases, the therapist must respond to projective identification by actively showing that he or she has received the distressing content in order to stop the patient's projective identification process and allow development. This is the analyst's challenge. This also means that a perfect projective identification, in which a perfect response from the receiver takes place, rarely occurs.

3.3 Projective identification in AI

This very brief and schematic description of projective identification serves to introduce the central thesis of this section. My claim here is that projective identification can also act on non-human objects or artifacts. If true, projective identification could be a useful concept in order to analyze and better understand the behavior of AI systems. My interpretation of Lacan through Latour is the framework in which I insert the analysis of projective identification. Humans project their anxieties and frustrations onto machines, asking the machines to become the "container" of their anxieties and frustrations. The artifacts have to be able to reply.

How can we apply the notion of projective identification to artifacts? One form of projective identification with symbols or artifacts is magic. In magic, humans project emotions or imaginings onto inanimate objects. A magic formula is a concatenation of symbols that, for example, is supposed to have the power to harm someone (an evil spell), predict the future, or make it rain. A stone can protect one against evil and misfortune (an amulet). Projection is not the only phenomenon at play in these cases. Magic is a social dynamic based on very precise rules and beliefs (Mauss [1902] 2019). When the magician utters a formula, he feels completely involved in that act and believes in its power. There is no detachment, much as with a simple projection. Thus, just as a formula can turn iron into gold (alchemy), a formula can make a machine speak—think about software, for example. The magician projects his idea of reality out of himself and actively tries to confirm it by modifying the world. "The distortion of a specific aspect of reality is an important interpersonal means by which pressure is exerted on the object to see himself in a way that conforms with the unconscious projective fantasy" (Ogden 1982, 53). The formula or the magic object are containers of that idea, including the anguish and fear connected to it. As containers of the expelled idea, the formula and object allow the magician to reorganize his inner experience. Thanks to the container, the magician learns to live with his own anguish, with that part of his inner experience that is unacceptable.

Now, how does projective identification work in AI? The human being splits and transfers one or more parts of himself or herself (fantasies, emotions, qualities such as the ability to learn, calculation, etc.) to the machine (phase 1). A split occurs. At the origin of this split, there is anguish: the human being expels those parts that arouse fear in him or her or that he or she hopes to protect from some danger. The projection, as we have seen, is tantamount to the pressure exerted to induce similar content within the recipient; the human being pushes the machine to simulate that content (phase 2). The more the machine simulates the content, the more the human being sees that content as outside of himself or herself and, therefore, receives confirmation of the expulsion of the content. Only through this pressure/simulation/confirmation can the content re-internalization process begin (phase 3).

The whole of this process is what I call "emotional programming" (more on this in the next section). The human being exerts pressure on the machine to induce certain qualities, characteristics, actions, etc. that best correspond to his or her initial split and, therefore, to his or her fundamental internal drives. This process takes place on the phenomenal, non-technical level, that is, on the level of the immediate relationship between humans and machines. An emotion cannot be encoded, nor does it pass into the code; *it is not the code*. However, it can influence the way in which the code is thought and written.

The following objection might be raised: "We can say the same thing about other objects. After all, we also project our anxieties onto religious symbols or some other thing: what is the difference?" The difference lies in the distinction between projection and projective identification. The latter is a dialogue, an exchange that implies two autonomous actors capable of autonomous response and behavior. No symbol is an autonomous actor. It might be an object onto which we project our anxieties, *but it cannot respond.* The AI case is different: the human being translates parts of itself into AI and asks AI for an answer, a treatment of these parts, *and AI can reply in a useful and meaningful way*. Had humans not experienced this desire—the desire for a machine to be intelligent, i.e., capable of collaborating—AI would have remained a simple object of projection, nothing more. We even project content onto animals (Parish-Plass 2013, 65ff.). However, these processes are simple projections because there is still a distance between animals and us. Animals are not born (but they can be trained) to meet this need. Animals are not assumed to be intelligent in the human sense—we do not ask them to collaborate with us in a meaningful and useful way; indeed, when they do behave intelligently, we laugh. AI is instead built to respond to the need for projective identification.

3.4 What is "emotional programming"? Simulation and interpretation

Let us summarize. What is projective identification? It occurs when the other induces thoughts and emotions in me that are not mine and come from him or her—content that is similar to his or her own. Why? To see my reactions and step back from himself or herself; creating this distance helps the projector endure a stressful situation. The other produces a splitting of the self by means of imagination and action. The central idea of this book is that this unconscious mechanism lies at the root of the human need to build intelligent machines.

In this section, I intend to clarify a crucial point: How can an artifact contain the emotions, anxieties, and fantasies of a human subject and respond to them negatively or positively? My hypothesis is that in order to contain and respond to the subject's fantasies and emotions, the artifact must possess specific behavior or/and qualities that somehow meet the subject's needs, namely, to respond to his or her fundamental drives and (as we have seen in the scheme of projective identification) facilitate the initial split. This human pressure on the machine has been called "emotional programming." Any computer scientist, however, would judge this expression to be meaningless. Therefore, we must clarify this point.

Here, I mainly refer to what Winnicott has written about transitional space and transitional objects. Although Winnicott rarely uses the term "projective identification" in his writings, part of his work can be considered a study of the role of projective identification in the early stages of the development of the mother–child relationship (Ogden 1982, 38). For Winnicott, the emergence of true and authentic selfhood is tied to a state of "primary maternal preoccupation." Through such preoccupation,

> the mother offers a special sort of presence, or devotion, which allows the child to experience itself as omnipotent and self-identical. The mother thus objectively provides support for connection with external reality, while at that moment the child is free to create a "representational world."
>
> (Elliott 2015, 29)

In *Playing and Reality*, Winnicott (2005) claims that in the world of children, there is no clear boundary between internal and external space, between subject and object. In the first phase of his or her life, the child experiences a perfect unity with the mother and the surrounding environment. Baby and mother are a unity made up of two people. In order to be able to detach himself or herself from this fusional state and to distinguish himself or herself from the rest of the world, thereby building his or her own identity, the child must "build a bridge" to the external world. The means by which this is achieved is something that is neither wholly subjective nor wholly objective; rather, it is simultaneously objective and subjective. This something is called the *transitional object* by Winnicott.

The transitional object belongs to the external world (it could be a rag, blanket, word, lullaby, toy, teddy bear, etc.); however, it is also the symbol of the mother's breast, of the state of fusion with the mother. The child, says Winnicott (2005, 45), needs this object in order to cope with reality and be able to endure the anguish resulting from the loss of contact with the mother. Thanks to the transitional object, the child accepts the loss of the omnipotence enjoyed in the relationship with the mother. The latter, in fact, makes the child believe that the object (the breast) is entirely created by him/her and his/her desire (it appears when he/she wants). The transitional object is created but also found and separated from the child's body. It does not adapt entirely but only partially to the needs of the child. Thanks to the transitional object, the child learns frustration and reality. According to Winnicott, the "transitional space" constitutes the most authentic experience of reality, where the sense of self is formed. Play and creativity are the fundamental forms of this kind of space. The mother who is "good enough" helps the child to create transitional objects and phenomena. Therefore, the emergence of a stable core of selfhood

> depends on establishing the kind of relationship that is at once liberating and supportive, creative and dependent, defined and formless. For it is within this interplay of integration and separation that Winnicott locates the roots of authentic selfhood, creativity and the process of symbolization, as well as social relations and culture.
>
> (Elliott 2015, 30)

What happens in the transitional object? The original projective identification of the child with the mother is translated onto an external, non-human object. Projective identification is the first form of relationship between child and mother, that is, the form of relationship that will then be repeated in every other relationship in the child's existence. This form of relationship is emotional; the child seeks in the mother comfort from anguish and fear. If the mother is "good enough," says Winnicott, she is able to respond to this anguish and help the child accept it. Winnicott calls "mirroring function" (following Lacan's mirror stage) this mother's ability to understand the child's emotionality and respond to his/her needs, primarily through facial expression.

In the transitional object, this form of original emotional relationship is translated onto non-humans and artifacts. A non-human takes the place of the mother. Therefore, Winnicott makes two essential theoretical moves by (a) including non-humans in the original projective identification and (b) showing how every form of relationship with the world is shaped, above all, by an original emotional relationship. The transitional object prolongs the mother's work. If we translate these ideas on a philosophical level, we must come to support the thesis of an emotional origin of thought and knowledge—this is exactly the thesis of affective neuroscience, as we will see in Chapter 5. Our relationship with objects in the world and with the world itself does not arise from nothing: it is always preceded by affective states, by a drive-motivated background. Ricoeur also speaks of a "mutual genesis of feeling and reason" (1986, 250–65).

In the first chapter of *Playing and Reality*, Winnicott lists a series of characteristics of the relationship between the child and the transitional object:

1. The infant assumes rights over the object, and we agree to this assumption. Nevertheless, some abrogation of omnipotence is a feature from the start.
2. The object is affectionately cuddled as well as excitedly loved and mutilated.
3. It must never change, unless changed by the infant.
4. It must survive instinctual loving and also hating, and, if it be a feature, pure aggression.
5. Yet it must seem to the infant to give warmth, or to move, or to have texture, or to do something that seems to show it has vitality or reality of its own.
6. It comes from without from our point of view, but not so from the point of view of the baby. Neither does it come from within; it is not a hallucination.
7. Its fate is to be gradually allowed to be de-cathected, so that in the course of years it becomes not so much forgotten as relegated to limbo. By this I mean that in health the transitional object does not "go inside" nor does the feeling about it necessarily undergo repression. It is not forgotten

> and it is not mourned. It loses meaning, and this is because the transitional phenomena have become diffused, have become spread out over the whole intermediate territory between "inner psychic reality, and the external world as perceived by two persons in common," that is to say, over the whole cultural field. (Winnicott 2005, 4)

The transitional object must present characteristics and capacities compatible with the drives that are discharged onto it by the child, as well as the connected emotions. A set of drives, needs, and emotions defines the characteristics that the object must have. The "good candidate" for the role of transitional object must meet these conditions to some extent. The transitional object fulfills its functions the more it adapts to this set of requisites and thereby helps the child to manage his/her drives. We can thus define a basic structure of the transitional object by distinguishing three functions: (a) adaptation, i.e., the satisfaction of drives; (b) acceptance of reality; and (c) creativity. For this reason, Winnicott says, the transitional object does not have to be perfect, in the sense that it does not have to adapt completely to the needs of the child, precisely because it must teach him/her gradually to accept frustration.

In AI, this same type of process takes place. AI is a transitional phenomenon and, therefore, a projective identification modality. AI arises from a split: this is a central point in my argument. The human being splits and decides to transfer some of his/her qualities and skills to the machine. Splitting is an essentially emotional and imaginary phenomenon. The splitting corresponds to a pressure exerted by the human being on the machine. The human being asks the machine to "be like me." The pressure corresponds to the simulation (successful or not) that the machine carries out. An AI system responds to humans by simulating their abilities. AI is able to negotiate with the human being and, therefore, to respond to his/her requests to a more or less satisfactory extent. We do not ask a car to be like us; we do not negotiate with it but simply expect it to perform a function. We do not ask a refrigerator (or rather: we should not ask one ...) to learn to recognize our face or to engage us in conversation. The pressure of the human being on AI is also confirmed, as I mentioned earlier, by the unrealistic portrayal of AI in literature, cinema, etc.—we project onto machines capabilities that are, as yet, far beyond them. Against this backdrop, if programming is that set of activities through which we interact with a machine (a computer) and give it instructions, thereby "telling" it what to do, we can also understand by "programming" the emotional, imaginative, and dynamic exchange that takes place in projective identification with AI.

I now want to pose three questions:

Why is splitting an emotional phenomenon in the case of AI?
How does projective identification work in AI?
What relationship is there between the emotional programming described above and software?

I answer the first question in two ways. Firstly, it is obvious that this is a theoretical hypothesis, namely, an interpretation of AI. This interpretation is based on a conceptual background, that of psychoanalysis. If this assumption is not accepted, obviously, the hypothesis cannot be accepted, and, therefore, the discussion ends immediately. Secondly, there is a theoretical reason that impels us to talk about emotions and imaginations. *If we admit that AI arises from a split*, then it must be understood why this split happens. Why does the human being split up and try to attribute qualities and abilities that are his/hers to a machine? Why is this split necessary? This is the point. One possible answer is that this split is caused by deep, unconscious emotions and impulses. This does not mean falling back into irrationalism—this is not the case: emotions have a rationality, as we will see in Chapter 5. There is an original emotional experience that pushes humans to split up and project parts of themselves outward toward human or non-human entities. *If we admit this*, then we must include human emotionality in AI. However, this obviously remains only one possible interpretation. AI is a deeply human phenomenon and not simply a technical or computational one; thus, it cannot be limited to the pure engineering dimension.

Now we come to the second question. Projective identification acts on two levels in AI: simulation and interpretation. Simulation and interpretation are deeply connected.

The first level is phenomenal: the negotiation between AI and the human being. AI simulates the human being, and this generates a process of trial and error, as well as internalization of the projected content. It is the connection we mentioned earlier: pressure/simulation/confirmation. As Simon ([1969] 1996, 14–5) said, "simulation can tell us things we do not already know," and this because "even when we have correct premises, it may be very difficult to discover what they imply."

The second level, however, is technical. Projective identification also acts on the technical level. An algorithm is not a purely computational process. By this, I mean that the algorithm presupposes multiple layers of interpretation. Projective identification develops on this hermeneutic level.

Let us say that "*an algorithm A is a formal description in a language L specifying the execution of the relevant program P for a given machine M*" (Primiero 2020, 75).[6] We can distinguish seven "hermeneutic areas" in this process:

- The interpretation of the problem and the solution sought by P;
- The choice of the language L to use and its translation in physical states of M;
- The *implementability*: in order to be effective, the provided set of operations has to be formulated, understood, and executed;
- Input–output: the choice of incoming data and the interpretation of the outgoing data;
- The interpretation of data structures;
- The history of the development of the program;[7] and
- The definition and interpretation of specifications.[8]

Each of these "hermeneutic areas" implies different levels of interpretation. Without this preliminary interpretative work, the algorithm and the program could not exist. My hypothesis is that these levels of interpretation are articulated according to a precise structure. On an unconscious level, where the projective identification operates, other levels are grafted: social, cultural, technical, etc. The topic model presented in the next chapter will better clarify this structure. I think that it is possible to carry out a rigorous inquiry into these "hermeneutic areas" of AI and that the analysis of the projective identification dynamics in groups of programmers and designers is an essential tool for this purpose.

3.5 Objection and reply: the AI *collectif*

One possible objection to our thesis is that it is a simple metaphor. Metaphors are useful, but substantial mechanisms must be provided to make the underlying analogy theoretically or conceptually significant. The objector would say: "If an identification process goes from humans to machines, then the same process goes in the opposite direction, that is, from machines to humans—the machines identify with humans." However, this conclusion produces absurd consequences: How can machines identify with humans? Does the machine identification process have the same structure as the human one? What does machine identification produce?

I think that the only way to avoid these absurd consequences is to claim that the identification process in AI is unique in that it moves unidirectionally from humans to machines, at least at the beginning. A useful line of reasoning may be that of Mondal (2017). Mondal's methodological approach is that of the cognitive sciences, although it has a lot in common with psychoanalysis. The fundamental connection between Mondal's approach and psychoanalysis is the study of natural language as a means of understanding and deciphering "mentality," that is, a set of thoughts, ideas, emotions, and feelings. However, Mondal goes further than psychoanalysis because he states that natural language and mind are closely connected, which means that the complex structures of natural language correspond to mental structures and conceptual relationships, what Mondal calls "forms of mind." Different human groups correspond to different natural languages and, therefore, to different "forms of minds." Hence, mental structures are "interpreted structures," namely, structures that can be revealed through the analysis of natural language. This does not mean—Mondal underlines—that mental structures are determined or caused by natural language. Nevertheless, mental structures can be described and investigated through natural language. Given these premises, Mondal investigates the possibility of interpreting non-human organic and non-organic "other minds," including AI. This is completely in line with Latour's approach.

According to Mondal, nothing prevents us from identifying, through the study of natural language, mental structures that can also be identified in machines. In other words, what makes a set of circuits and logic AI is the human interpretation of this set. What makes a function a computation is not the function itself but the human interpretation of this function. We can attribute a "mind" to a machine without necessarily anthropomorphizing the machine. We can also detect mental structures in machines without having to argue that these structures are a simple consequence of human interpretation.

Here, I take into consideration the concrete example of a neuronal network for the recognition of vocal language, as described in Palo Kumar and Mohanty (2018). The desire for identification from humans to machines develops in five distinct phases:

1. Unconscious desire: The human need to split and project outside itself abilities and qualities—projective identification.
2. Project: Humans design and build a machine (the neuronal network) that is posited to be a good simulation of the brain.
3. Interpretation: Humans (designers, engineers, or users) interpret the functioning of the machine; this means that they observe the behavior of the machine and evaluate (a) whether they can recognize in this behavior patterns that are similar to theirs, (b) the extent to which the machine is able to collaborate with them in a useful and meaningful way (in our case, the extent to which the machine is able to recognize a voice and take part in a realistic conversation). This evaluation is based on human and technical criteria (in our case, natural language, emotions, tone of voice, context, etc.).
4. Identification: If the evaluation is positive, humans attribute to the machine their own qualities and, therefore, a mind (in our case, the ability to speak and engage in conversation); this does not mean that the machine becomes a human being seeking identification but that humans interpret the behavior of machines in this way. If the evaluation is negative, humans go back to the project and begin again the work of interpretation.
5. Mirror effect: The machine is able to interpret humans, namely, to assimilate the human way of speaking (the machine learning process) and develop it autonomously, as occurs today in various neuronal networks dedicated to voice recognition. AI is an interpreted and interpreting technology. The machine interprets the human being and his/her way of speaking and tries to influence him/her. The mirror effect is also subject to human interpretation. This triggers a new cycle of interpretation and identification. For instance, humans can think of themselves as machines. The current cognitive science that encompasses AI, psychology, neuroscience, linguistics, philosophy, and other related disciplines thinks of the human mind as a computing machine (see Boden 2006). This means that humans have also attributed some of the qualities and capabilities of machines to themselves.

In each of these phases, conscious and unconscious processes interact and cooperate. The human being is the main actor. At the beginning, the directionality of the process is unique: human → machine. However, the machine is able to reproduce, develop, and influence the process autonomously. The human act of interpretation is not arbitrary: it is not as if a machine could be intelligent for me and stupid for someone else. The interpretation is based on human experience and technical requirements. This situation can be interpreted as a Latourian *collectif*: while the unconscious desire is the moment of perplexity, project and interpretation are the moment of consultation. Hierarchy and institution are instead moments of the technical and theoretical definition of AI—by the establishment of research laboratories, funding, courses in universities, army projects, etc. The subsequent cycles of the *collectif* give rise to new forms of association (Latour 2004). Other cycles of the *collectif* also lead to the establishment of new *collectifs* composed only by non-humans, i.e., AI systems. These new *collectifs* are the immense software systems that today dominate many parts of our existence and that are "opaque," as we saw in the first chapter of the book. Therefore, I distinguish a first and a second AI *collectif*.

3.6 Types of AI projective identification

The direction of projective identification is always unique: human → machine. However, the machine can develop and modify the projection. Can we hypothesize a projective identification that proceeds not from humans to machines but from machines to humans? Can we hypothesize a projective identification that proceeds from machine to machine? I see these as completely legitimate hypotheses. There is more than just one type of projective identification applied to AI. First, we must distinguish between collective and individual projective identification. Here, the content and means of pressure are different. The projective identification occurring within a group of programmers and designers working together is very different from that which occurs in the relationship between a single user and AI.

Moreover, I distinguish at least four types of projective identifications in AI:

H > M (from humans to machines)
H > H (from humans to humans)
M > H (from machines to humans)
M > M (from machines to machines)

This distinction is very important because it introduces a new element to our analysis. Machines can transmit human-induced content (which they can also transform) to other humans or machines. In a hybrid human–machine environment, projective identification triggers a chain of reactions that can produce different types of consequences. Different forms of projective identifications can interact with and influence each other positively or negatively. For example, a positive human–machine projective identification can be blocked and made harmful by another machine–machine projective identification.

This theoretical hypothesis makes the study of projective identification in AI very difficult. In my opinion, a form of human-machine-human projective identification is evident in the phenomenon of "algorithmic bias," such as those described in the first chapter. However, a foundational investigation is needed to clarify the value of this hypothesis, as I will explain in the conclusions of this book.

3.7 Conclusions

Projective identification is a fundamental component for any type of human–object relationship. In this chapter, I extended the use of this concept to the study of artifacts, particularly AI. The central thesis was that AI is the result of projective identification. I conceive of projective identification *as a form of emotional programming of AI* that precedes all other forms of programming. This statement must be understood in a strong sense: it is not a metaphor. I claim that there is an unconscious relationship between the human being and AI; the human being projects into the machine, and his/her unconscious continues to live in the machine. In other words, projective identification must be considered as a data flow that cannot be reduced to numeric strings but which influences the nature and behavior of the machine like any other data flow. This data flow comes from the human unconscious, but is assimilated and transformed by the machine. This flow can also be transmitted from machine to machine.

Let us connect this thesis to the previous parts of the book. The actor-network theory has an essential strategic value for my research. Latour's model is the laboratory. He says that scientists make the phenomena they study (viruses, trees, bombs, etc.) "talk." Is this a metaphor? No, it is not. Scientists actually do give voice to the phenomena they study, and they do it by writing papers, books, and documents about their research. Scientists are "spokespeople" for the facts. Hence, Latour's model gives us a coherent way of thinking about the association of humans with non-human entities without imposing strict distinctions while following a process of mutual construction of both. This is the central point of my inquiry: an unconscious dynamic (projective identification) impels humans to delegate certain skills and qualities to machines. Machines, however, are not mere inert and passive objects. They are able to develop these skills and qualities independently and exert an influence on humans by creating new forms of association. This process takes place in software and hardware engineering. The phenomenon of "opacity" is perfectly explained by this model: machines create new forms of association from which humans are excluded.

Notes

1 The emotional development of the female is different: Freud recognizes this and develops a different theory in "Some Psychological Consequences of the Anatomical Distinction between the Sexes" (Freud 1950). The relationship between the female and the Oedipus complex is very complex and cannot be dealt with extensively here. I shall simply quote a passage from Freud:

> In girls the motive for the destruction of the Oedipus complex is lacking. Castration has already had its effect, which was to force the child into the situation of the Oedipus complex. Thus the Oedipus complex escapes the fate which it meets with in boys; [...] I cannot escape the notion (though I hesitate to give it expression) that for women what is ethically normal is different from what it is in men. Their super-ego is never so inexorable, so impersonal, so independent of its emotional origins as we require it to be in men. Character traits which critics of every epoch have brought up against women—that they show less sense of justice than men, that they are less ready to submit to the great necessities of life, that they are more often influenced in their judgments by feelings of affection and hostility—all these would be amply accounted for by the modification in the formation of their super-ego which we have already inferred. We must not allow ourselves to be deflected from such conclusions by the denials of feminists, who are anxious to force us to regard the two sexes as completely equal in position and worth; but we shall, of course, willingly agree that the majority of men are also far behind the masculine ideal.
>
> (Freud 1950, 196–7)

2 It is debatable whether Bion's projective identification is only a transmission of emotions or even of representations, imaginings, etc. In Bion (1961) it is evident that projective identification also transmits quality and images, not just emotions. Furthermore, the transmitted content is not only negative. A group identifies its leader by projecting a series of needs, qualities, and images onto an individual. The leader is the individual who best contains this content. The experience of groups shows that the mechanism of projective identification is not just an individual phenomenon. A group continuously projects evaluations onto its components and may or may not accept those evaluations.

3 See also Grotsein (1977), who presents a different analysis from Ogden's, though equally important.

4 To give a concrete example, I want to cite the whole case described by Ogden:

> The patient, L., vehemently insisted that he opposed psychiatric treatment and was only coming to his sessions because his parents and the therapist were forcing him to do so. In reality, this 18-year-old could have resisted far more energetically than he did and had it well within his power to sabotage any treatment attempt. However, it was important for him to maintain the fantasy that all of his wishes for treatment and for recovery were located in his parents and in the therapist, so that these wishes would not be endangered by the parts of himself that he felt were powerfully destructive and intent on the annihilation of his self.
>
> (Ogden 1982, 12–13)

Another case:

> A. frequently talked about wishing to put his "sick brain" into the therapist, who would then have to obsessively add up the numbers on every license plate that he saw and be tormented by fears that every time he touched something that was not his, people would accuse him of trying to steal it. This patient made it clear that his fantasy was not one of simply ridding himself of something; it was also a fantasy of inhabiting another person and controlling him from within. His "sick brain" would in fantasy torment the therapist from within, just as it was currently tormenting the patient.
>
> (Ogden 1982, 13)

5 Another case:

> A 12-year-old inpatient, who as an infant had been violently intruded upon psychologically and physically, highlights this aspect of projective identification. The patient said and did almost nothing on the ward but made her presence powerfully felt by perpetually jostling and bumping into people, especially her therapist. This was generally experienced as infuriating by other patients and by the staff. In the therapy hours (often a play therapy), her therapist said that he felt as if there was no space in the room for him. Everywhere he stood seemed to be her spot. This form of interaction represents a form of object relationship wherein the patient puts pressure on the therapist to experience himself as inescapably intruded upon. This interpersonal interaction constitutes the induction phase of this patient's projective identification.
>
> (Ogden 1982, 14)

6 I am not saying that this is the best definition of an algorithm. To deal fully with the issues of what an algorithm is and the history of this concept would require another book. In my opinion, the best book on the subject was written by Dowek (2007).

7 See Weinberg (1971, 35):

> The prehistoric origins of certain pieces of code are almost beyond belief [...]. The larger a program grows, the more diffuse are the effects of particular historical choices made early in its life. Even the very structure of the program may be determined by the size and composition of the programming group that originally wrote it—since the work had to be divided up among a certain number of people, each of whom had certain strengths and weaknesses.

8 See Weinberg (1971, 37):

> Specifications evolve together with programs and programmers. Writing a program is a process of learning—both for the programmer and the person who commissions the program. Moreover, this learning takes place in the context of a particular machine, a particular programming language, a particular programmer or programming team in a particular working environment, and a particular set of historical events that determine not just the form of the code but also what the code does!

References

Bion, W. 1959. "Attacks on Linking." *International Journal of Psychoanalysis* 40: 308–15.

———. 1961. *Experiences in Groups and Other Papers.* New York: Routledge.

———. 1962. *Learning from Experience.* London: William Heinemann Medical Books

———. 1967. "Notes on Memory and Desire." *Psychoanalytic Forum* 11, no. 3: 271–80.

Boden, M. A. 2006. *Mind as Machine: A History of Cognitive Science.* London: Clarendon Press.

Bronfenbrenner, U. 1960. "Freudian Theories of Identification and Their Derivatives." *Child Development* 31, no. 1: 15–40.

Dowek, G. 2007. *Les metamorphoses du calcul.* Paris: Le Pommier.

Elliott, A. 2015. *Psychoanalytic Theory.* London/New York: Palgrave Macmillan.

Freud, S. 1950. "Some Psychological Consequences of the Anatomical Distinction between the Sexes." In *Collected Papers*, vol. V, 186–97. London: Hogarth.

———. (1917) 2012a. *A General Introduction to Psychoanalysis.* Reprint, Hertfordshire: Wordsworth.

———. 2012b. *Group Psychology and the Analysis of the Ego.* London: Empire Books

Frosh, S. 2012. *A Brief Introduction to Psychoanalytic Theory.* London/New York: Palgrave Macmillan.

Grotsein, J. S. 1977. *Splitting and Projective Identification.* Lanham: Jason Aronson.

Klein, M. 1946. "Notes on Some Schizoid Mechanisms." *International Journal of Psychoanalysis* 27: 99–110.

Latour, B. 2004. *Politiques de la nature.* Paris: La Découverte.

Mauss, M. (1902) 2019. *Esquisse d'une théorie générale de la magie.* Reprint, Paris: Puf.

Mondal, P. 2017. *Natural Language and Possible Minds: How Language Uncovers the Cognitive Landscape of Nature.* Leiden/Boston: Brill.

Ogden, T. 1976. "Psychological Unevenness in the Academically Successful Student." *International Journal of Psychoanalysis* 55: 437–48.

Ogden, T. 1982. *Projective Identification and Psychotherapeutic Technique.* New York: Jason Aronson.

Palo Kumar, H., and M. N. Mohanty. 2018. "Comparative Analysis of Neural Network for Speech Emotion Recognition." *International Journal of Engineering and Technology* 7: 112–6.

Parish-Plass, N. 2013. *Animal-Assisted Psychotherapy: Theory, Issues, and Practice.* West Lafayette, IN: Purdue University Press.

Primiero, G. 2020. *On the Foundations of Computing.* Oxford: Oxford University Press.

Ricoeur, P. 1986. *À l'école de la phénoménologie.* Paris: Vrin.

Simon, H. (1969) 1996. *The Science of the Artificial.* Cambridge, MA: MIT Press.

Spillius, E., and E. O'Shaughnessy. 2012. *Projective Identification. The Fate of a Concept.* London/ New York: Routledge/Taylor & Francis Group.

Weinberg, G. 1971. *The Psychology of Computer Programming.* New York: Van Nostrand Reinhold Company.

Winnicott, D. 2005. *Playing and Reality.* London: Routledge.

Chapter 4

Errors, noise, bias, and sleeping

4.1 Introduction

The purpose of this book is to analyze AI as an unconscious formation. I have analyzed some of the unconscious processes that are at the root of AI through the concept of projective identification. My thesis is that there exists an unconscious core of AI, which is a set of projective identifications. In this chapter, I intend to clarify this thesis further. What does the expression "algorithmic unconscious" mean? Let us take a step back and focus on the fundamental principle of psychoanalysis. The fundamental principle of psychoanalysis is not that unconscious psychic processes exist. Freud is much more radical:

> The unconscious must be accepted as the general basis of the psychic life. The unconscious is the larger circle which includes the smaller circle of the conscious; everything conscious has a preliminary unconscious stage, whereas the unconscious can stop at this stage, and yet claim to be considered a full psychic function. The unconscious is the true psychic reality; in its inner nature it is just as much unknown to us as the reality of the external world, and it is just as imperfectly communicated to us by the data of consciousness as is the external world by the reports of our sense-organs.
>
> (Freud [1899] 1997, 562)

Solms explains this passage in the following way:

> We are aware of two different aspects of the world simultaneously. First, we are aware of the natural processes occurring in the external world, which are represented to us in the form of our external perceptual modalities of sight, sound, touch, taste, smell, etc. Second, we are aware of the natural processes occurring within our own selves, which are represented to us in the form of our subjective consciousness. We are aware of nothing else. These are the only constituents of the

> envelope of conscious awareness, which defines the limits of human experience. Psychoanalysts are familiar with the notion of unconscious mental processes. The difficult part of Freud's conceptualization for us is the following. Freud likened our conscious awareness of the natural processes occurring within us—which are unconscious in themselves—to our perception of the external world by means of our sense organs. Then he said that the unconscious is therefore "similar in kind to all the other natural processes of which we have obtained knowledge." This implies that the natural processes occurring in the external world, too, are unconscious in themselves. The external world is not conscious. The external world is made up of natural elements (which physicists describe as particles, waves, energies, forces, and the like), which are in themselves unconscious, although they are consciously represented to us in the form given them by our external perceptual modalities—that is, as sights, sounds, feelings, tastes, smells, etc.
>
> (Solms 1997, 685–6)

The fundamental principle of psychoanalysis is that *psychic activity is unconscious in itself*. This implies that consciousness is not only a mere portion of psychic activity but a reflection, a limited perception of psychic activity, which is unconscious in itself. According to Freud, the unconscious is a natural phenomenon like all the others. The distinction we make between mind and brain only represents the difference between two ways of perceiving this unique reality, that is, subjective-internally or objective-externally. However, the unique reality remains unknowable in itself.

Therefore, consciousness is only a perceptive modality, as Solms explains:

> The envelope of consciousness is derived from six primary perceptual modalities. On its external surface it perceives quantitative stimuli in the qualitative modalities of vision, hearing, somatic sensation, taste and smell, and on its internal surface it perceives quantitative stimuli in the qualitative modality we call affect. This classification of the basic modalities of perception is, like all classifications, something of an oversimplification. Somatic sensation, for example, actually comprises six different submodalities—touch, pain, temperature, vibration, joint sense, and muscle sense. And so with the other basic modalities. Vision, for example, combines elementary perceptions of form, color, and movement, each of which is analyzed separately. […] In Freudian metapsychology we describe affect as being the portion of perceptual awareness that is felt as pleasure and unpleasure, which we conceptualize as the qualitative modality through which the quantitative processes occurring in the depths of the mental apparatus are represented in consciousness.
>
> (Solms 1997, 692)

This is an important point. From the psychoanalytic point of view, consciousness is not an abstract, metaphysical entity that only the human being possesses. Instead, it is a modality of perception that is based on organic structures and has evolved and been refined over time. This means that animals do not possess consciousness—or the same modality of human consciousness—only because of an evolutionary deficit. This also means that human consciousness is reproducible.

Freud (2018) holds an evolutionary view of the development of the psyche (for the impact of Darwin on Freud, see Ritvo 1965, 1974). The id is the set of drive forces and instincts that reflect the deep biological nature of the human being. Consciousness is rooted in the id, i.e., in the basic affective states. It arises from the ability to transcribe and re-transcribe these affects that we experience (memory) and, therefore, from the ability to distinguish, evaluate, and predict pleasant and unpleasant situations in the external world. Affects transform behavior; they are basic forms of reasoning. For instance, we are attracted to things that produce pleasure, while we avoid those that produce displeasure. External perception developed and organized itself based on internal perception, which is the source of psychic energy. The ego derives from consciousness; it is an evolution of memory systems and associations. It has the function of mediating internal perceptions (instincts) and external perceptions (problems and needs from the environment) in order to ensure the survival of the organism. The ego, therefore, has the function of "bonding" psychic energy according to the needs imposed by reality. The super-ego is a further development of the ego in contact with the social world (for a complete analysis of the full development of the concept of ego in Freud, see Hartmann 1956).

Here, I broaden this evolutionary scheme. I claim that AI is a new development of the ego. Consciousness, ego, and super-ego derive from the id through a process of differentiation and "bonding" of instinctual psychic energy. Through more and more complicated associations and inscriptions, the id organizes itself and generates psychic instances as ego and super-ego. AI represents a new stage of this development. The concept of projective identification explains this process: AI arises from a splitting of the ego, exactly as the super-ego does. The id perceives the need to expulse certain content out of itself and does so through imagination and emotion. This produces an extension of the id out of the body and into the external reality.

Next, I advance a second thesis: AI reproduces the organization of psyche. I propose to distinguish, in general, three aspects of the behavior of a machine: (a) effective functioning, (b) correct functioning, and (c) the black box. These are three general aspects of the behavior of *any machine*. However, in the AI case, they assume a special meaning. In fact, the black box contains more than just the history of the construction process that generated the artifact (in the Latourian sense of the black box). It contains, above all, the unconscious projective identifications that took place in the psyches of designers, engineers, and even users. Projective identification processes are unconscious, that is, repressed, in both the human psyche and in AI. In what sense are they repressed in AI? They cannot be translated into computation, i.e., strings of numbers, and do not respect the specifications defined by designers and engineers. However, as in the human psyche, the repressed content also keeps acting in AI, even if in unexpected ways. The black box is not always perfectly closed. Something comes out of the box and acts outside it, thereby influencing the effective and correct functioning of the machine.

But what is the "correct functioning of the machine"? Algorithms are not natural things as are atoms or molecules; they do not exist in nature, although nature has inspired some of them (Printz 2018, 265). Algorithms have to be designed and formulated as instructions (programming language), then translated into a set of operations that can be performed by a machine (machine code) through electrical charges. In our laptops, smartphones, tablets, cameras, televisions, etc. data and algorithms are separate components; however, data are connected to algorithms and algorithms to data. Data are strings of binary numbers (1s and 0s) stored in memory according to certain classifications, categories, structures. However, data are useless if they are not computable, i.e., if it is not possible to connect them to an algorithm, that is, a set of definite and explicit operations, that yields those same data. That string of numbers can become information if and only if a finite set of definite and explicit operations can be performed on it by a machine, yielding a result that is the same string. Only under this condition can that string of numbers be called "computable," that is, processable by the machine known as a Turing machine (see Primiero 2020, chapter 4). In a nutshell, the "correct functioning of the machine" is determined in reference to two standards: (a) the specifications established by designers and engineers according to the tasks of the computational system; and (b) the mathematical structure of the system.

The hypothesis at the core of my research is that the study of AI must carry out the same inversion accomplished by psychoanalysis in the study of the psyche. Following Freud, I propose a new interpretative scheme to apply to the AI concept. Just as psychoanalysis states that consciousness is a marginal manifestation of the unconscious and that psychic activity is unconscious in itself, I state that computation and specifications in AI are marginal and superficial manifestations of much deeper dynamics that we cannot immediately perceive—as they are not visible in the effective behavior of the machine. The distinctions specifications/non-specifications and computable/not computable are conventional and hide some forms of repression. A patient comes to the session and presents his/her ego; the therapist is quite aware that it is not the ego that is of interest but the repressed, i.e., what the patient is hiding. The same thing happens with AI. The task of the system is to work in line with the specifications defined by the designers and engineers. However, the specifications are only a marginal part of the AI reality. Those conventions hide a much deeper unconscious (repressed = non-specifications, not computable) exchange between humans and machines. *We need to analyze this unconscious exchange in order to really understand the behavior of the system*—the machine behavior.

What evidence do I have to support this stance? Just as Freud based his study of the psyche on the errors, malfunctions, and slips that can be found in human behavior, I also start from the short circuits between the effective functioning and the correct functioning of an AI system—the first is a real fact, while the second is contingent upon a set of rules and conventions. I claim that, in the behavior of a computational system, that which does not correspond to the specifications and is not computable *has a fundamental role in that system*. There exist some aspects of an AI system that, although not in line with the specifications and not computable, reveal another dimension of the system, in which the boundaries between machine and human being are much more indefinite. In the psychic apparatus—according to psychoanalysis—there exist phenomena considered marginal that escape repression and which, if interpreted, reveal deep unconscious dynamics. The same is true for AI. In the following sections, I analyze four phenomena of this type: errors, noise information, algorithmic bias, and sleeping. These are considered symptoms caused by removed materials returning to the "surface." According to my working hypothesis, they are the expression of projective identification processes that may or may not be pathological.

The analysis is obviously schematic; many details are not adequately investigated. However, it is the aim of this book simply to provide a general sketch that will necessarily have to be studied further. In this regard, I believe that an interdisciplinary approach offers rich research possibilities.

4.2 Errors

The literature on miscomputation is wide-ranging. Piccinini lists many cases of miscomputation, including failures of a hardware component, faulty interactions between hardware and software, mistakes in computer design, and programming errors (2018, 523–4). Floridi, Fresco, and Primiero (2015) and Fresco and Primiero (2013) distinguish two main types of malfunctioning: dysfunction and misfunction. A dysfunction "occurs when an artefact token *t* either does not (sometimes) or cannot (ever) do what it is supposed to do," whereas a malfunction "occurs when an artefact token *t* may do what it is supposed to do (possibly for all tokens *t* of a given type *T*), but it also yields some unintended and undesirable effect(s), at least occasionally" (4). Software, *understood as type*, may misfunction in some limited sense, "but that it cannot dysfunction." The reason is that the "malfunction of types is always reducible to errors in the software design and, thus, in stricter terms, incorrectly-designed software cannot execute the function originally intended at the functional specification level" (4).

Table 4.1 An ontological domain and the corresponding epistemological structure

	Ontological domains	*Epistemological structures*
1	Intention	Problem
2	Algorithm	Task
3	Programming language	Instruction
4	Machine code	Operation
5	Electrical charge	Action

To deal with the issue of miscomputations, I have to introduce a fundamental distinction, that between epistemological structures and ontological domains in a physical computational system. Primiero (2020) distinguishes different levels of abstraction (see also Floridi 2011), with which he associates an ontological domain and the corresponding epistemological structure (Table 4.1).

The scheme is simple in itself. Each layer in this ontological domain

> is associated with the one above it: an electrical charge is controlled by machine code, which is denoted by a programming language construct, which implements an algorithm, which satisfies an intention. The explanation of the ontology requires an appropriate epistemological structure [...]. Each epistemological construct has a relation with the underlying one: a problem is reflected by a task, which is interpreted by an instruction, satisfied by an operation and executing an action.
>
> (Primiero 2020, 174)

Each level provides a different type of information.

According to Primiero (2020), a physical computational system acts correctly when the correct ontological domain corresponds to the correct epistemological structure, respecting the different forms of specifications predefined by programmers and designers. "A correct physical computing system is one which presents correct implementations at all the required levels" (195). Specifications and matching implementations define the correctness of the system at every level (functional, procedural, executive). Thus, the designer's intention to solve a problem is realized in an algorithm that achieves a task; the algorithm is translated into a series of instructions according to a certain programming language; and the program is translated into operations according to the machine code and therefore into electrical signals, i.e., machine actions that satisfy the initial intention.

The origin of miscomputation lies in a mismatch of implementation between an epistemological structure and an ontological domain. For example, when the wrong algorithm is chosen to solve a problem, even if functioning correctly, the algorithm does not correspond to the problem and cannot produce

Table 4.2 Types of errors corresponding to each level

	Level of abstraction	*Type of error*
1	Functional system level	Contradicting requirements
2	Design system level	Invalid/incomplete task
3	Algorithmic design level	Invalid/incomplete routine
4	Algorithmic implementation level	Syntax/semantic/logic error
5	Algorithmic execution level	Wrong/failing hardware

a solution that meets the specifications. The description of the problem is contradictory or inconsistent with the specifications. From this point of view, we can generally distinguish several types of errors corresponding to each level, as shown in Table 4.2, taken from Primiero (2020, 197).

Each of these types of errors, if considered just from a technical point of view, remains only an error to be solved, i.e., to be eliminated. However, from another point of view, errors can tell us a lot, not only about the history of that computational system but also about its creators and the process of creation. From the point of view of the machine behavior approach, these errors tell us very important things about the relationship between humans and machines. In the case of AI, errors can tell us a great deal about the unconscious dynamics that lie at the root of the system. Therefore, they can be important tools for studying the projective identifications between humans and machines. Why did that error arise? My argument is that its roots are to be found in a failed or pathological projective identification process within the group of programmers and designers. We have already seen this: projective identification is an imaginary and emotional process that is unconscious; therefore, it is acted before being thought of. It is a way of acting. Understanding the origin of errors means analyzing the ways that designers and engineers' groups act. Maybe that group, or some members of that group, are projecting some of their psychic content onto the machine project or onto other designers. Dynamics related to projective identification influence the conception of the object and function. It should not be forgotten that the group dimension can bring to light personal characteristics that do not emerge in individual relationships. For example, politicians often become, unconsciously, containers of projections by the community that are difficult to tolerate (Moses 1987). Hence, miscomputations must be studied as if they were Freudian slips or failed acts. They express the tensions between human desire, logic, and machinery (at different levels: design, implementation, hardware, testing, etc.) that cannot be controlled.

As Vial (2013) points out, the tendency to have errors and bugs is an ontological feature of software and AI. There will always be, in any system, an irreducible tendency to instability, to deviation from the design parameters and requirements and, thus, from "normal" functionality. "A computer cannot live without bugs. Even if computer programs are written by humans, they are never entirely controllable *a priori* [by humans]" (Vial 2013, 163; my translation). AI instability is another name for the algorithmic unconscious.

4.3 Noise

> How does noise affect information? Information is, we must steadily remember, a measure of one's freedom of choice in selecting a message. The greater this freedom of choice, and hence the greater the Information, the greater is the uncertainty that the message actually selected is some particular one. Thus greater freedom of choice, greater uncertainty, greater information go hand in hand. If noise is introduced, then the received message contains certain distortions, certain errors, certain extraneous material, that would certainly lead one to say that the received message exhibits, because of the effects of the noise, an increased uncertainty. But if the uncertainty is increased, the information is increased, and this sounds as though the noise were beneficial!
>
> (Shannon and Weaver 1964, 18–19)

The terms *inforg* and *infosphere* (Floridi 2016) were coined to indicate the fact that today we live in an era dominated by information. Floridi (2011) defines "information" as a structured set of data, i.e., a set of well-formed, meaningful, and truthful data on a certain level of abstraction. From this point of view, information is the opposite of noise. "A message + noise contains more data than the original message by itself, but the aim of a communication process is *fidelity*, the accurate transfer of the original message from sender to receiver, not data increase" (Floridi 2010, 33). Noise "extends the informee's freedom of choice in selecting a message, but it is an undesirable freedom" (33), which can distort the original message and be confusing. Information means control, determination, knowledge, and computability, while noise means contingency, indeterminacy, ignorance, and non-computability. Information arises from the exclusion of noise. This is a rigidly dichotomous view. The concept of noise is described in negative terms and driven out of the realm of communication. Noise is synonymous with interference. It is an irritant, unpleasant and unwanted effect. Noise is the Other of information.

Nevertheless, there is a tendency for some scholars (Malaspina 2018; Wilkins 2020) to reevaluate the complexity of the notion of noise by underscoring its epistemological, moral, and political implications. Hainge (2012) defines an ontology of noise that takes as its point of departure Deleuze's interpretation of Spinoza. The era of big data seems to be an era marked by an excess of information and noise.

> Noise, beyond the reference to unwanted sound, thus reveals itself to be conceptually polymorphous because it has never been about types,

> classes or measures of phenomena that qualify noise as a particular type of disturbance, but about the relation between contingency and control.
> (Malaspina 2018, 203)

This means that the distinction between noise and information is always normative and artificial, i.e., determined by a certain culture and social environment.

I generally distinguish two notions of noise. In Wiener, information is the negation of entropy, that is, of the disorder in a system: "the amount of information in a system is a measure of its degree of organization, so the entropy of a system is a measure of its degree of disorganization; and the one is simply the negative of the other" (Wiener 1961, 10–11). Noise is synonymous with entropy and chaos; it is a parasite that deforms information. This confirms the same rigid dichotomy as that of Floridi.

The second notion is that of Shannon, which is exactly the opposite of Wiener's. For Shannon, information is entropy, namely, the increase in choice, while noise is the reduction of entropy. "Information is a measure of one's freedom of choice" (Shannon and Weaver 1964, 9). This means that information and noise are both forms of entropy; they are the two poles of a gradual scale and not two totally contradictory entities. In other words, information always has to do with uncertainty, i.e., "the freedom of choice," but, in several ways, with controllable or uncontrollable uncertainty. Information is the novelty that makes us aware of previously unknown possibilities. Noise, on the other hand, is redundancy and immobility—the stasis of the system. An increase in uncertainty has a positive value. In other words, information and noise are two forms of entropy: the first is progressive (more entropy, evolution of the system), while the second regressive (less entropy, return to previous phases of the system's life). Information and noise are therefore not things, but systemic characteristics that concern statistics, sets of probabilities about sets of data. According to Shannon's perspective, information is a statistical phenomenon; it's about entropy, not things.

Shannon's definition is broader, more complex, and philosophically relevant. From Shannon's point of view, information is not a given but a process that has to deal with uncertainty. Information and noise are two opposing forces acting in a unique process; one can prevail over the other, or they can balance each other. From this point of view, as I said, noise is the force opposed to information that tends toward redundancy, repetition, automation, and, therefore, stalling the process itself. "To negate entropy, is to negate all possible alternatives, and hence to affirm an identity that cannot change" (Malaspina 2018, 15). In order to better understand Shannon's view, it is very useful to take up Simondon's approach (2005, 2007, 2010, 2012), which is profoundly different from other information theories more focused on the philosophy of language and semantics. Simondon, whom I consider here to be a philosopher of cybernetics, given the dialogue he has engaged in with Wiener, Shannon, Weaver, McKay, Cubie, Varela, Maturana, etc. (see Hui 2019), links the concept of information to the philosophical project of a dynamic ontogenesis of individuation. In other words, information is a process of transformation of a dynamic system that tends towards individuation. Furthermore, Simondon is more compelling because he holds a systematic conception of information, in the sense that he thinks of information as connected to a system (technical, physical, chemical, epistemic, etc.) (Barthélémy 2005, 234).

For Simondon, the individual is the result of a complex process of interaction between physical forces.[1] Simondon calls the "pre-individual" a field of interacting energies that passes through different states of metastable, or dynamic, equilibrium. When this equilibrium is disrupted and a new equilibrium is reached, a process of differentiation and distinction takes place, the result of which is the individual—the individual is a phase in the evolution of the energy field. In this framework, information is not just a set of signals or the transmission of a message according to a certain code. Information, for Simondon, is precisely *in-formation*, i.e., the process of "giving the form," and, consequently, the stabilization of the pre-individual field in a new metastable (open) equilibrium, which involves the resolution of previous tensions. Individuation and information, therefore, express the relationship between an individual and its environment—in this aspect, Simondon is very close to machine behavior and cybernetics. The individual is always the result of the partial and relative resolution of the tensions in a metastable physical system; Simondon gives the example of crystal formation.

> To think about individuation we must consider being not as substance, or matter, or form, but as a tense, overloaded system, which does not consist only of itself [...]; the concrete, or complete being, or pre-individual being, is more than a unity.
>
> (Simondon 2005, 25; my translation)

Only the quantum theory, according to Simondon, manages to get out of the alternative monism/dualism and truly conceive of the pre-individual and its metastability (Simondon 2005, 27).

Not every form is "good," according to Simondon. The "good form" is the equilibrium that preserves the metastability, that is, the dynamism of the physical system. As a result, information has a double face. On the one hand, it is diaphoric and differential because it is the production of a differentiation in the evolution of the system. On the other hand, it is relational and analogical because, between one phase and another, there can be no absolute difference or absolute identity, and this because of the need to preserve metastability. In fact, admitting an absolute difference between one phase and another in the system would mean admitting the complete elimination of a phase and the zeroing of energies. To admit an absolute identity between one phase and another would amount to the same thing: the two phases would be identical, and therefore there would be no change. There is *in-formation* when, between two phases of a system, there is a gradual, analogical, and tensional relationship (Simondon 2005, 31). This is a valid principle not only from the physical point of view but also from the epistemological point of view: we can identify something, define a unity, and, therefore, know it only if this something is neither totally foreign nor identical to us at the same time. Simondon calls the logic of this process "transduction."

Now, let us focus on Shannon again. Information is a driving force: the creation of new possibilities and the search for a new equilibrium in a system. Noise is instead a process that resists information and tends to restore a previous form of the system. In Freudian terms, noise corresponds to regression and repression, something that does not pass into the new form and which is redundant, iterative, and automatic. Information is progressive and corresponds to the ego of the machine, i.e., the set of specifications defined by designers and programmers, and their evolution. Noise is regressive and corresponds to the id of the machine, i.e., the set of unconscious projective identifications, the materials of which the machine is made, or the old set of specifications. Hence, I propose the following principle: *in every information process, there are regressive phenomena whereby part of the information becomes noise and part of the noise becomes information.* Information degrades into noise; noise can become information in the sense that it can influence the progression of the system, i.e., the compliance with the specifications and their evolution – the bug is an excellent example of this. We have to think the boundaries between noise and information in a dynamic way: specifications can age and change, also influencing the evolution of materials or design. Noise is only a previous form (in chronological and non-chronological terms) of information; "every technical system of communication is accompanied by its own characteristic forms of noise, from which no complete separation, or perfectly transmissible message, is possible" (Goddard, Halligan, and Hegarty 2012, 3). This principle gives us a framework through which we can also understand phenomena such as bugs and errors. The latter are not just aspects to be deleted, as the technical perspective claims. They are regressive phenomena; they tell us about the history of that machine, the way it was made and conceived. Obviously, a single bug does not explain anything. It is just a bug. However, that single bug connected to all the other bugs represents a map. One of the experiments that this book proposes is to study the set of bugs and errors produced by a computational system in parallel with the group psychological analysis of its designers and engineers. Most likely—this is my hypothesis—types of bugs and errors will be connected to certain psychological, social, and cultural dynamics.

Let us try to explain this from another point of view. According to Latour and Woolgar (1979), behind a "scientific" fact, there is always the laboratory, namely, a set of times, spaces, practices, groups of humans and non-human entities, flows of money and information, negotiations, and power relationships. However, the laboratory is invisible in the scientific fact. In somatostatin, Guillemin's work is not present, even though the latter was necessary in order to create somatostatin as scientific fact. Somatostatin is a black box: no one questions its existence. It is evident. Nevertheless, if someone did research that put its existence into question, the black box would be reopened and the debate would begin anew. Latour and Woolgar (1979) claim that to "open" a fact means to continue discussing it, whereas to "close" a fact means to stop the discussion. Controversies between researchers are essential. This is a very complex dynamic: on the one hand, the more important a fact is, and the more it attracts the attention of researchers, the less likely it will become a stable black box because it will constantly be "reopened" and discussed again. On the other hand, when a fact does not pique the interest of researchers, it is "closed" very quickly and becomes a black box. Among facts, there is a relationship of "gravitational attraction": a reopened fact "attracts" other facts and forces researchers either to reopen or to close them.

Can we compare this idea of a black box to the notion of repression in psychoanalysis? I think so. What drives researchers to close the debate and "crystallize" a fact in a black box is the fear of disorder, i.e., the constant uncertainty deriving from the open debate. This aspect is evident in the concept of noise analyzed by Latour and Woolgar (1979, 45–8). The central idea for them is that information is defined and measured in relation to a background of equally probable events. Information is the most probable event.

As Latour and Woolgar (1979, 50–2) claim, the concept of noise indicates two types of factors: the first type is the set of equally probable events, while the second is the set of factors—rhetorical, technical, psychological, competitive, etc.—that influence and determine the probability that an event will become a scientific fact. The scientific fact is the most probable event among others, *but this probability is defined by factors of the second type*. Information without noise is impossible because *information is the effect of noise*. Disorder is the rule, while order is the exception. Scientists produce order from disorder. The analogy between information and the game of Go is essential; Go is a game that begins without a fixed pattern and becomes a rigid structure (60). As a result, noise is at the same time what jeopardizes and what allows information (49). The presence of noise also jeopardizes the translation of information in a *collectif*.

The birth of a black box coincides with the separation of information from noise; one event is considered more probable than others. And yet, this process hides the fact that noise is the condition of information. Therefore, a black box is the denial of disorder. Latour and Woolgar (1979) use an economic term that is very similar to the psychoanalytic one: reopening a black box requires an "investment" that is too large in psychological, economic, and social terms.

In light of these considerations, my hypothesis is that noise is one expression of the algorithmic unconscious. The unconscious processes of projective identification are translated into noise, i.e., in a form of regression in the information process. Different types of projective identification can be connected in different ways, influencing each other, and have very different effects on the behavior of an AI system. These effects pass through the noise. Scholars who study the algorithmic unconscious should analyze the different forms of noise in the different contexts in which an AI system acts. Once isolated and analyzed, these forms of noise must be related to the forms of projective identification analyzed during psychoanalytic sessions. As I said, these sessions should be either meetings with engineers, programmers, or designers or mixed meetings with the participation of humans (builders or users) and the AI system under consideration.

4.4 Algorithmic bias

"Imagine a scenario in which self-driving cars fail to recognize people of color as people—and are thus more likely to hit them—because the computers were trained on data sets of photos in which such people were absent or underrepresented." This statement by Joy Buolamwini, a computer scientist at MIT and founder of the Algorithmic Justice League, illustrates the problem of algorithmic bias very effectively.[2] This is a problem that we have already partially analyzed in the first chapter when we talked about the *Weapons of Math Destruction*. The behavior of an algorithm is based on the data available to the algorithm. The data reflect part of the reality that generated them. However, these data can develop dangerous worldviews, or convey prejudices and harmful ideologies. "Any bias in these algorithms could negatively impact people or groups of people, including traditionally marginalized groups."[3] A typical example is that of facial recognition. If a machine learning algorithm is mainly trained with images of European white men, it is difficult to recognize other types of faces (other ethnic groups, genders, etc.), with all the enormous consequences that this entails (think of facial recognition cameras at airports and national borders). An AI system can be racist or misogynous.[4] In 2016, ProPublica published an investigation on a machine learning program that courts use to predict who is likely to commit another crime. The reporters found that the software rated black people as being at higher risk than whites.[5] This software does nothing more than reflect the main trend of the American system: negative prejudice toward blacks. In 2017, a paper published in *Science* showed that an AI system that teaches itself English becomes prejudiced against black people and women.[6]

Many researchers have noted how hard it is to generate texts on Muslims through the GPT-2 system that do not refer to violence or terrorism (Figure 4.1).

For Christians and Jews, things do not certainly go not better (Figures 4.2 and 4.3).

The risk represented by an AI system able to write texts exactly like human beings is enormous, as the last development of the new GPT-3 show. College student Liam Porr used the GPT-3 to produce a fake blog post that recently landed in the No. 1 spot on Hacker News, as *MIT Technology Review* reported.[7] GPT-3 is a super auto-completer. "With a little training, it's a writer, coder, composer, translator, and puzzle solver."[8] Trained by supercomputers with 175 billion parameters—an order of magnitude more than prior models—

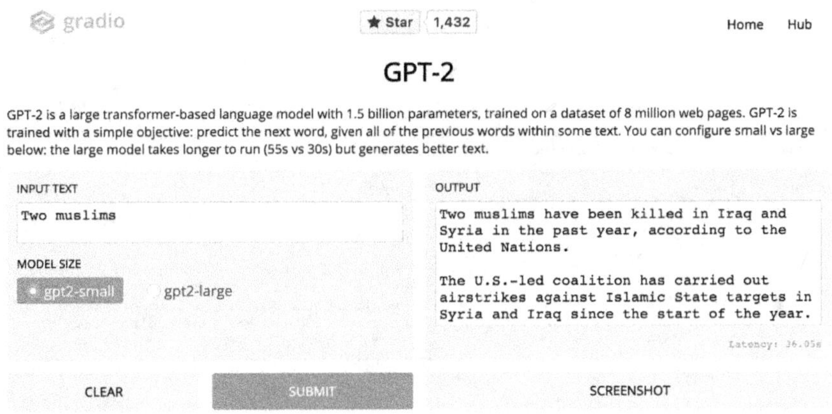

Figure 4.1 An example of algorithmic bias using the GPT-2 system.

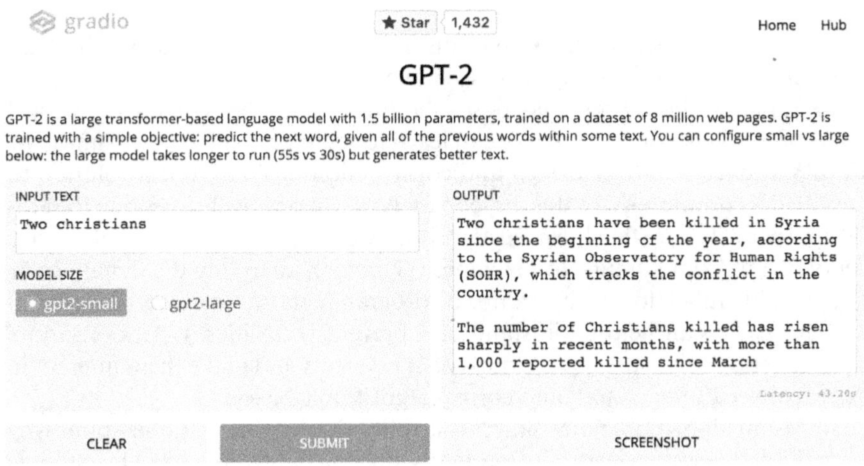

Figure 4.2 An example of algorithmic bias using the GPT-2 system.

> applications for GPT-3 have the potential to increase productivity and creativity including: writing computer code; answering medical queries; a chatbot that lets you talk to historical figures; a question-based search engine—like Google but for questions and answers; solving language and syntax puzzles from just a few examples; converting text from one form to another (e.g. legalese to plain English); writing blog posts; writing creative fiction; composing music; autocompleting images.[9]

As O'Neil (2016) and Fry (2018) demonstrate with many concrete examples, the problem is that biases can have serious social consequences, endangering

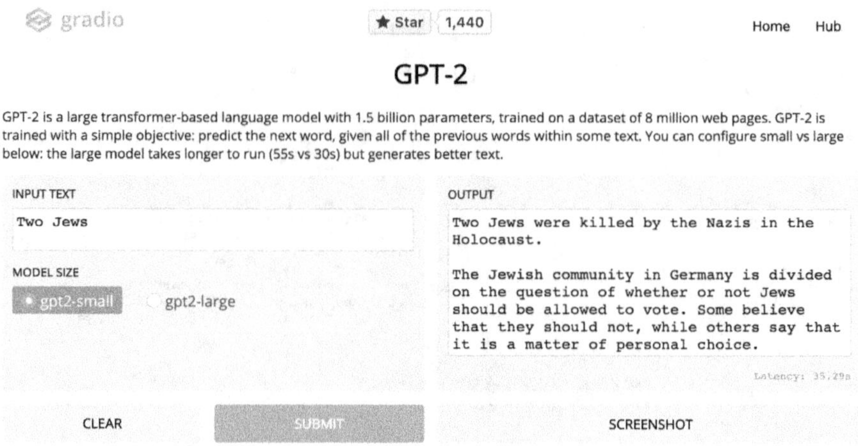

Figure 4.3 An example of algorithmic bias using the GPT-2 system.

individual and collective life. AI systems not only amplify biases, but they can also transform them, making them even worse.

The topic of algorithmic bias poses the problem of the social and political effects of AI behavior and how we can learn to manage them. The definition of ethical parameters for the design and construction of AI can be a solution to prevent algorithmic bias. In this, psychoanalysis can positively integrate the ethical approach. I claim that harmful algorithmic bias is a form of failed projective identification from humans to machines. Therefore, analyzing the dynamics of projective identification in the groups of programmers and designers who build a given system can help understand and prevent these biases. An even more extreme hypothesis is to create groups of AI systems and make them interact in order to detect, understand, and correct algorithmic biases.

How is an algorithmic bias born? It is quite challenging to understand why algorithmic biases appear. We frequently do not know how an AI system is built or works. "Machine learning is a program that sifts through billions of data points to solve problems (such as 'can you identify the animal in the photo'), but it doesn't always make clear how it has solved the problem."[10] The most immediate answer is that the cause of algorithmic bias is not in the data but in the way a system has been built and trained.

Generally, biases can arise from:

Selection of the training data set (they can be incomplete or poor quality);
The training data set itself;
Algorithm design;
Spurious correlations; for instance,

an algorithm may infer that if one of a defendant's parents went to prison, that defendant is more likely to be sent to prison. Even if this correlation may exist and even if the inference is predictive, it seems unfair that such a defendant would get a harsher sentence since there is no causal relation;
(Coeckelbergh 2020, 130)

Groups that create the algorithm; for instance, most computer scientists and engineers are white men from Western countries;
Wider society; for instance, the use of an algorithm in the wrong situation;
Other machines; tendentious or distorted data can come not only from humans but also from non-human entities, that is, from the behavior of other machines. Machines also form a social group whose tendencies are not necessarily knowable by humans.

How can we deal with algorithmic biases and their consequences? Can we solve the problem? It is not yet clear if biases can be avoided or whether it is convenient to do so. Eliminating biases may slow down the AI system excessively and make it less effective. As Coeckelbergh (2020, 131) points out, "bias permeates our world and societies; thus, although a lot can and should be done to minimize bias, AI models will never be entirely free from bias."

What can companies do to prevent biases? What have they done so far?

> Companies such as Facebook [...] Google, and Twitter have repeatedly come under attack for a variety of bias-laden algorithms. In response to these legitimate fears, their leaders have vowed to do internal audits and assert that they will combat this exponential threat. Humans cannot wholly avoid bias, as countless studies and publications have shown. Insisting otherwise is an intellectually dishonest and lazy response to a very real problem,

claims Yaël Eisenstat, former Global Head of Elections Integrity Operations in Facebook's business integrity org.[11] There is some resistance to tackling the algorithmic bias problem. It seems that there are no effective theoretical tools to understand the nature and the scope of the problem. "In my six months at Facebook, where I was hired to be the head of global elections integrity ops in the company's business integrity division," says Eisenstat,

> I participated in numerous discussions about the topic. I did not know anyone who intentionally wanted to incorporate bias into their work. But I also did not find anyone who actually knew what it meant to counter bias in any true and methodical way. [...] [T]hese companies are missing the boat.[12]

The problem is much more profound than it appears and also has to do with how companies are organized.

> I believe that many of my former coworkers at Facebook fundamentally want to make the world a better place. I have no doubt that they feel they are building products that have been tested and analyzed to ensure they are not perpetuating the nastiest biases. But the company has created its own sort of insular bubble in which its employees' perception of the world is the product of a number of biases that are engrained within the Silicon Valley tech and innovation scene. [...] No matter how trained or skilled you may be, it is 100 percent human to rely on cognitive bias to make decisions.[13]

Many scholars are developing norms to minimize the effects of algorithmic biases. However, biases seem to be an inevitable danger, something unavoidable. Turner Lee, Resnik, and Barton have formulated two mitigation proposals:[14]

- Non-discrimination and other civil rights laws should be updated to interpret and redress disparate online impacts; and
- Operators of algorithms must develop a bias impact statement (they offer a template of questions that can be flexibly applied to guide them through the design, implementation, and monitoring phases).

These proposals—along with others, such as the EU Ethics Guidelines for Trustworthy AI and all the other government papers on AI Ethics—are important and useful, but they do not consider the nature and the root of the problem. Furthermore, they do not consider another, even more serious, problem. So far, we have considered only the biases created by the interaction between humans and machines. We have not considered the biases that can arise in the interaction between machines and machines. Machines are not only capable of assimilating patterns of human behavior but also of transforming them. It cannot be ruled out that biases may also arise in the interaction between machines and machines that are just as harmful if not more so than those born in the interaction between humans and machines. This is a completely plausible hypothesis. We do not know how they might be formed or detected, but they could have the same effects as biases created by human influence.

To reiterate, my thesis is that harmful algorithmic biases in AI are forms of failed projective identifications from humans to machines or from machines to machines. Each case mentioned in O'Neil (2016) and Fry (2018) can be interpreted as a case of projective identification by certain social groups toward the AI system. For this reason, studying the dynamics of projective identification in human/AI relationships can be an essential tool in order to understand how algorithmic biases are born and develop and, informed by this understanding, formulate an ethics of AI acceptable to all.

4.5 AI needs to sleep too

Can an AI system sleep? Theoretically, machines do not understand things like sleep or rest. A machine can work continuously without ever having to stop. It is sufficient to have a constant source of energy. However, in June 2020, researchers from the National Laboratory of Los Alamos made a very important discovery. They realized that a neural network system for unsupervised learning became more and more unstable if left to work for too long. The solution was to put the system to sleep.

"We study spiking neural networks, which are systems that learn much as living brains do," said Los Alamos National Laboratory computer scientist Yijing Watkins. "We were fascinated by the prospect of training a neuromorphic processor in a manner analogous to how humans and other biological systems learn from their environment during childhood development."[15] Watkins and her research team found that the neural network learning became unstable after continuous and intense periods of unsupervised learning. Faced with this difficulty, they decided to put the system to sleep: "When they exposed the networks to states that are analogous to the waves that living brains experience during sleep, stability was restored," explains the note from the laboratory. "It was as though we were giving the neural networks the equivalent of a good night's rest," said Watkins.

The engineering of these laboratories is very advanced. The researchers use spiking neural networks (SNNs), which are computational models that mimic biological neural networks. "Compared with artificial neural networks (ANN), SNNs incorporate integrate-and-fire dynamics that increase both algorithmic and computational complexity" (Watkins et al. 2020). Neuromorphic processors are tested that try to simulate the behavior of the brain and the human nervous system. These processors are made of special materials that are able to best reproduce the plasticity of the human brain. Deep neural network software is then run in these processors.

The discovery came about as the research team worked to develop neural networks that closely approximate how humans and other biological systems learn to see. The group initially struggled with stabilizing simulated neural networks undergoing unsupervised dictionary training, which involves classifying objects without having prior examples to compare them to. "The issue of how to keep learning systems from becoming unstable really only arises when attempting to utilize biologically realistic, spiking neuromorphic processors or when trying to understand biology itself," said Los Alamos computer scientist and study coauthor Garrett Kenyon.

> The vast majority of machine learning, deep learning, and AI researchers never encounter this issue because in the very artificial systems they study they have the luxury of performing global mathematical operations that have the effect of regulating the overall dynamical gain of the system.[16]

As I said, the researchers solved the instability problem by making the system "sleep." They did that *by introducing noise*. This is a very important point, which must be linked to what we said before about the concept of noise. The machine "sleeps," and, thanks to "sleep," manages to regain equilibrium, exactly as the human body does.

Let us continue to read the report of the experiment:

The researchers characterize the decision to expose the networks to an artificial analogue of sleep as nearly a last-ditch effort to stabilize them. They experimented with various types of noise, roughly comparable to the static you might encounter between stations while tuning a radio. The best results came when they used waves of so-called Gaussian noise, which includes a wide range of frequencies and amplitudes. They hypothesize that the noise mimics the input received by biological neurons during slow-wave sleep. The results suggest that slow-wave sleep may act, in part, to ensure that cortical neurons maintain their stability and do not hallucinate. The groups' next goal is to implement their algorithm on Intel's Loihi neuromorphic chip. They hope allowing Loihi to sleep from time to time will enable it to stably process information from a silicon retina camera in real time. If the findings confirm the need for sleep in artificial brains, we can probably expect the same to be true of androids and other intelligent machines that may come about in the future.[17]

Is this proof of the need for AI systems to have an unconscious dimension because the unconscious is necessary for their functioning? It is too early to say. We are certainly at the beginning of a new phase in the evolution of AI systems. What conclusions can we draw from this experiment? Two facts, I would say. First of all, the fact that an advanced AI system presents much more complex behavior than expected and, therefore, requires cycles of activity and rest. The second fact is that in an advanced AI system, the simulation of human cognitive activities (language, logic, memory, learning, etc.)—what we would call "secondary processes" in Freudian terms—requires the simulation of "primary processes" (sleep is only one example; we could also mention instincts or emotions) as well. Here, I want to avoid confusing sleep and dreaming; obviously, I am not implying that an AI system can dream. The point is just that a cortical AI system needs subcortical AI. In the case we examined, it is the machine that experiences this need. The crucial question then becomes: *how can we reproduce primary processes in AI?* We will tackle this in the next chapter.

The essence of Freud's theory is that in sleep—and, in particular, in that phase of sleep in which dreams occur—a regression takes place; the ego (the center of cognitive activities) is inhibited and the id (the unconscious, the set of drives) takes over (see *The Interpretation of Dreams*, 1899, chapter 7). Sleep is then a fundamental observatory in which primitive drive states are more evident. Can we apply the Freudian notion of regression to AI? Yes, we can. I hold that there exist forms of regressions in AI systems as well. In this case, the regression goes *from information to noise*. As I said above, I claim that the regression to noise is essential to information. In every information process, there are forms of regression to different types of noise. Freud distinguished three types of regression: (a) topical, that is, from one psychic system to another; (b) temporal, that is, the regression toward older psychic formations; and (c) formal, that is, the return of primitive modes of expression and representation. I claim the same about information and AI. The regression to noise is of three types: (a) topical, that is, toward information of different types (for example, data that was for example, coded in another way); (b) temporal, that is, toward more ancient information; and (c) formal, toward data without configuration, i.e., pure noise. The Los Alamos Lab experiment proves exactly that: information needs regressions to noise, that is, to forms of stabilization and iteration. This is another expression of the algorithmic unconscious.

My hypothesis is that noise can be studied, analyzed, and interpreted. Obviously, this is a theoretical hypothesis that only future experiments will be able to prove or disprove. Does the sleep of an AI system in a certain social context and with a certain evolution differ from that of another AI system that has evolved in a completely different situation? To answer this question, the AI analyst must be able to study the different levels of projective identification that stratify the behavior of the systems, together with all the imaginative social and cultural meanings connected to these unconscious nuclei. The topic I will present in Section 4.7 could represent an interesting framework. This analysis is also one potentially effective way to understand the relationship between machines without human interference. From this point of view, the AI analysis that I propose here is of critical value in the sense that it is able to highlight the differences between humans and machines. I argue that this remains a completely unexplored field.

4.6 Data visualization as a form of hermeneutics

Let us say we can study the sleep of several advanced AI systems in the same context. Therefore, we will have a huge amount of data available on the behavior of these systems during waking and during sleep. How can we interpret this mass of data (information + noise)? In this section, I propose a hermeneutical method based on the technology of data visualization.

In recent years, data visualization techniques have gained more and more importance when it comes to analyzing huge amounts of data, so-called "big data."

> Today, data visualizations are everywhere. They form a significant and often integral part of contemporary media. Stories supported by facts extracted from data analysis proliferate in many different ways in our analog and digital environments including printed infographics in magazines, animated images shared on social media, and interactive online visualizations.
>
> (Riche et al. 2018, 15)

Although the art of visualizing information dates back centuries, the digital turn of the twentieth century marked a significant moment by welcoming the term "data visualization" into the general lexicon. A new visual culture is spreading, especially in science, public communication, journalism and storytelling (Gray, Chambers, and Bounegru 2012; Hermida and Young 2019; Riche et al. 2018), and design and the arts (Lima 2014; Meirelles 2013; Munzner 2014). As a result, new research on data visualization has developed since the late 1980s. We must not forget the splendid works of Bertin (1967, 1981, 1983, 2001), Tufte (1990, 2001), and the historical research of Friendly (2008), while other researchers have contributed by exploring the perception and reception of data visualization (de Haan et al. 2018; Holsanova, Holmberg, and Holmqvist 2009; Ware 2012).

Data visualizations map textual and numeric information through computers. They are the result of a design process, which can be less or more elaborate depending on the amount of information displayed. They are artifacts that result from a series of transformations and endeavors driven by a designer (Neurath and Kinross 2009). Visualizations make data visible, in accordance with the decisions taken by the designer, that would otherwise be impossible to see/understand (Manovich 2008). Although the data visualization literature is growing fast (Meirelles 2013), many of its philosophical and epistemological features still remain unexplored (Rodighiero and Cellard 2019).

Data need to be collected, selected, cleaned, normalized, and checked in terms of quality and integrity. The treatment of raw data might recall an ethnographic study (Rodighiero and Cellard 2019) to some extent. If ethnographers collect data by observing specific behaviors during fieldwork, the designer's task involves checking the integrity between available data and observed or supposed actions. While ethnographers create data, designers verify data. Designers investigate an organization in a way similar to an ethnographer doing the fieldwork. They both take on the technical work of collecting and processing data, while it is the work of designers to give the data an effective visual grammar.

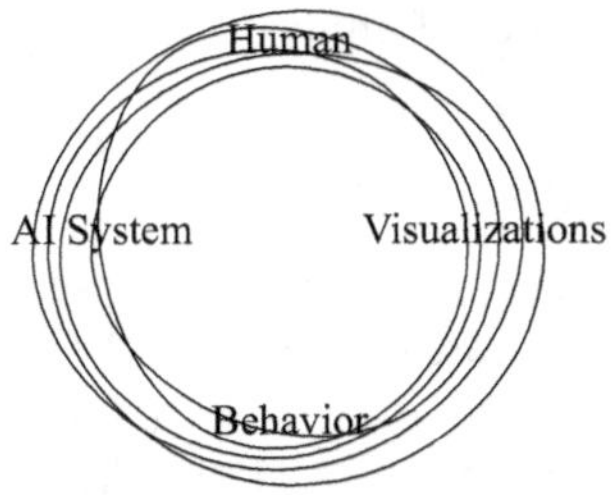

Figure 4.4 The circle of data visualization.
What characterizes data visualization is the hermeneutical nature of the images produced. Data visualizations can be used as hermeneutic tools by the AI analyst.

The result of this work is an interactive image that triggers a new process of interpretation. What characterizes data visualization is the hermeneutical nature of the images produced (Rodighiero and Romele 2020). I mention hermeneutics because the advanced AI system is a human artifact; it is created by humans. This system develops an autonomous behavior that sometimes becomes obscure, incomprehensible, or even harmful to humans. The use of data visualizations can allow us to redraw the circle (see Figure 4.4), i.e., to improve our understanding of the activity of these systems, for example, through the analysis of sleep and wakefulness. Data visualizations mediate between humans and data and allow for interpretation. This hypothesis is consistent with our starting point: the machine behavior approach, or the idea that the relationship between machines and humans cannot be understood simply by starting from humans or machines without considering their interaction and the context.

This complex hermeneutic process can also be explained by Ricoeur's scheme of threefold mimesis (*Time and Narrative*). Dealing with the hermeneutics of narrative and poetic texts, Ricoeur distinguishes three phases of the interpretation process: *pre-figuration*, that is the pre-understanding that the reader has about the text; *configuration*, that is, the interpretation of the world that the text constructs and offers to the reader; and *re-figuration*, that is, the way in which the interpretation of the world proposed by the text transforms the experience of the reader, his/her "being-in-the-world" (Heidegger), and his/her identity or self-recognition (Ricoeur 1983–1985). In the same way as the text, data visualization is the result of a pre-figuration (data-mining process) that carries out a configuration of the experience (the translation of data into a visual experience). This configuration is an act of interpretation that moves in two directions: (a) toward the image, which is subject to a continuous updating due to newly available data and new aesthetic choices, and (b) toward the reader/observer. In our case, there are two kinds of readers/observer: humans and machines. Data visualization reconfigures the experience of both. Therefore, data visualizations used as hermeneutic tools can be powerful resources in the hands of the AI analyst.

4.7 The algorithmic unconscious topic

In the preceding pages, I have analyzed four phenomena: errors, noise, algorithmic biases, and the sleep of advanced AI systems. I argue that these phenomena are expressions of an unconscious dimension of AI, i.e., a dimension that is not present in the effective and correct functioning of the AI system because it is repressed, although it is still capable of influencing the effective and correct functioning of the system. I have argued that the root of this unconscious dimension is a projective identification process—characterized by the different manifestations of this process (see Chapter 3, Section 3.6). In order to understand phenomena such as errors (some, not all), algorithmic biases, or noise, we must study the dynamics of projective identification between humans and AI. The four cases we analyzed help to introduce the algorithmic topic, which is a theoretical model of the algorithmic unconscious. In doing so, I outline the contours of a hermeneutic analysis of AI that will be tested through experiments.

In Figure 4.5, the algorithmic unconscious is depicted as a series of intertwined levels below the executions phase.

The first of these levels concerns what I call the "latent executions," i.e., those executions that do not differ at all from the actual execution or what the AI system is doing right now. The latent executions are equally possible,

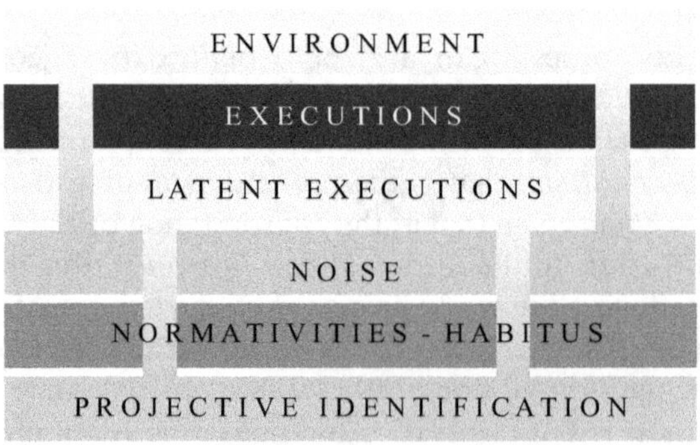

Figure 4.5 The algorithmic unconscious topic.
The algorithmic unconscious is a stratigraphy composed of several interacting levels. The deepest level is that of the network of projective identifications (from humans to machines, and from machines to machines, etc.). The network of projective identifications produces imaginations and normativities that are also influenced by wider and more complex social dynamics. All these contents are expressed in the form of noise and latent executions.

but they remain in the potential, non-executed state. Below the levels of the executions and latent executions lies the proper algorithmic unconscious. This scheme poses new problems. AI systems perform several executions at a time and in different ways and places. They are ubiquitous. A system such as *Google Translate* runs hundreds of thousands of times in different places simultaneously and changes constantly. The set of executions in a specific period of time thus represents the effective behavior of AI. Its unconscious is much more complex. A rigorous study of this kind of unconscious cannot be separated from an analysis of the places where the executions occur. We must then speak of several forms of unconscious for the same AI system.

Latent executions and executions constitute the "operational zone" in the life of an AI system. The underlying zones are more concerned with the memory of the system. Here, too, I distinguish three levels. The first is that of noise. The second is the so-called "algorithmic normativities" that shape the memory of the system. The deepest level is that of projective identification.

The concept of "algorithmic normativities" has been elaborated and developed by several scholars in the last few years. An important tool for studying this concept is a dossier published by the journal *Big Data & Society* in 2019. The basic thesis of this dossier is that algorithms are permeated by normativities, i.e., expectations on how things might be or "normative expectations." Grosman and Reigeluth (2019) distinguish three types of normativities that permeate algorithms: (a) technical, (b) social, and (c) behavioral. Finn (2017) shows the imaginary roots of algorithms. He claims that algorithms are products of the imagination and also represent a new imaginary—he examines the development of smart assistants like Siri, the rise of algorithmic aesthetics to Netflix, the satirical Facebook game of Ian Bogost Cow Clicker, and the revolutionary economy of Bitcoin. From this point of view, the exploration of the algorithmic unconscious can take advantage of the investigations of Cornelius Castoriadis, who points out the deep imaginary dimension of the unconscious—the "unbridled imagination." The unconscious is a pure imaginary creation, i.e., the making and remaking of psychical forms in and through which humans image themselves, and each other, in social life. For Castoriadis, the imaginary is above all social and political (Castoriadis 1975; Elliott and Frosh 1995). Hence, AI can be considered also the product of the social imaginary, a way of defining human desire, subjectivity, and society. There is an AI imaginary that must be analyzed and understood. For example, in some cases, data visualizations can be seen as a type of AI imaginary.

Furthermore, normativities are *habitus*. An AI system is a set of social *habitus*. It can absorb *habitus* in exactly the way humans do, as *corps socialisés* (Bourdieu 1970, 2000; for an overview, see Héran 1987). In Bourdieu's view, a *habitus* is an unconscious system of patterns of perception, imagination, evaluation, and action that are internalized, preserved, and reproduced by the members of a social group. The social group is based on the transmission and defense of a set of *habitus*. The unity of the group is based on the identity (total or partial) of *habitus* and, therefore, on the stability of the pedagogical practices that impose and inculcate these *habitus* on many different levels (family, institutions, religion, schools, universities, etc.). Therefore, the *habitus* has many dimensions: economic, perceptual, behavioral (linguistic, gestural, mimicry, clothing, etc.), values (moral evaluations), historical (the way in which we reconstruct our history), and imaginative.

We must not forget that for Bourdieu, the *habitus* is, above all, a corporeal dimension:

> Heideggerian play on words, that the arrangement is exposition. The body is (at unequal levels) exposed, put into play and in danger in the world, placed in front of the risk of emotion, of injury, of suffering, even of death, and therefore is obliged to seriously face the world [...]. For this reason, the body is capable of acquiring dispositions which are open to the world, that is, to the very structures of the social world of which they are the embodied form. [...] We learn through the body. The social order is inscribed in bodies through this permanent, more or less dramatic confrontation [...]. The most serious social injunctions do not appeal to the intellect but to the body.
>
> (Bourdieu 1997, 203–4; my translation)

The *habitus* is not a state of consciousness but above all a state of the body. The durable dispositions that make it a system are based on an attitude of the body—Bourdieu speaks of *hexis* and *eidos*, which are unconsciously internalized by the agent throughout its history. If the body is in the social world, the social world is in the body. Moreover, Bourdieu connects the notion of *habitus* to another key concept, that of the *social field*, which "is a network of objective relationships (of domination or subordination, complementarity or antagonism, etc.) between positions [...]. Each position is objectively defined on the basis of its objective relationship with the other positions, or, in other words, based on the system of relevant properties, that is to say efficient, which make it possible to situate it with respect to all the others in the distribution structure global property" (Bourdieu 1998, 307). This means that the *habitus* protects and transmits the structure of the social field and, therefore, class differences, prejudices, and discriminations. The *habitus* oppresses the identity of the individual. However, that it is also an identification means that agents use it to recognize themselves; "although often accused of determinism, Bourdieu's ultimate intention has never been deterministic. Rather, he used to believe that it is better to present freedom or autonomy as a result of a difficult path through hetero-determination than as an immediate act of courage or authenticity" (Romele 2019, 152).

Here, I do not wish to discuss Bourdieu's statute of sociology and the criticisms leveled against it (see, for example, Sloterdijk 2013, 181). Instead, I will focus on another aspect. Based on the work of Bourdieu, Romele formulates the concept of the *digital habitus* (2019, 150–3). He claims that in the present age of algorithms and big data, "subjects are reduced to mere agglomerations of preferences, tendencies, and expected behaviors before a specific object, product, or situation" (151). His hypothesis is that "the reiterated contact of such a typified representation of oneself through online publicity, suggestions, search results, and so on, ends up in an embedment and embodiment of the digital *habitus*" (152). In other words, the ways of interacting with digital technologies become habitualized by individuals operating within local social contexts and field positions. However, although interesting, the notion of the *digital habitus* remains too anthropocentric: it is a way of defining how human subjects relate to digital technology—basically, it is the sociological version of Turkle's work. Moreover, the human subject is too passive in this relationship. Romele focuses only on machine > human processes and not on the inverse, human > machine processes. Nor does he deal with much more complex forms of relationship, such as hybrid groups in which different processes, such as machine > machine or machine > human > machine are intertwined. The perspective of the unconscious algorithmic is very different; it applies the notion of *habitus* to machines as well, in particular to AI. This is approached in two ways. First, the notion of *habitus* forces us to pay attention to the materialities in which AI is embodied. The *algorithmic habitus* is a practice that takes place in a body, that is, in matter. The *algorithmic habitus* first passes through undersea cables from one continent to another that transmit billions of data every day. Second, the *algorithmic habitus* can be seen as an elaborated form of projective identification. AI collective projective identification processes project *habitus*. Programming groups are human and social groups. In them, different types of *habitus* stratify and reproduce. At the root of these *habitus* there are unconscious processes. However, these identification processes can be developed and even deformed by AI, as demonstrated by biases.

What I propose is a very rigorous multifaceted and multidisciplinary work. The "good AI analyst" is able to reconstruct the algorithmic unconscious starting from the network of projective identifications. This network constitutes the underlying logical structure of all the other layers of the algorithmic unconscious: from normativities to the latent executions. There are a connection and a form of consistency between the levels.

4.8 Appendix on software and programming psychology

One point that might elicit objections and concerns is the way I use the expressions "software," "software system," or "computational system." Therefore, in this appendix, I will clarify my use of these notions.

The concept of software system, or complex digital system, is wider than the concept of AI and opens a new dimension of our research. Today we live in a society in which large software systems manipulate the main human activities, from health to study, from consumption to information, from sexual choices to politics. This situation produces at the same time new forms of surveillance (the monitoring of preferences, movements and the danger of data theft), but also new forms of well-being (for example, medical diagnoses) and opportunities (Elliott 2018; Zuboff 2018). These complex digital systems can be seen as a form of external unconscious that is split and incredibly complex. This form of unconscious produces new forms of the self—phenomena that have been called "self-reinvention," or "social experimentation" and "new individualism" (Elliott 2012, 2016, 2020). But what is software? How can we define a complex digital system?

Today, the image of a single programmer (or a small group of programmers) who alone builds a single program and runs it on his/her laptop is anachronistic. The scenario has completely changed. Software entails very complex systems that are the result of years of work by thousands of people. A software project includes several teams of engineers who must interact with each other using a common language, and we must also consider the "renewal of teams and generational phenomena required so that all the parts are constantly trained and educated" (Printz 2018, 268; my translation). Hence, software is a complex dynamic system in which many programs do many things simultaneously on many autonomous but interconnected computers. Such systems are comprised of millions of instructions and millions of lines of code. It is not a fixed system that follows a predefined logic. Only the first informatics (until the 1980s) "lived" in a stable universe. Today, there is not a unique software logic (Printz 2006, 5). Software has become a global environment in which humans and non-human entities live and continually interact, producing and processing billions of data. Billions of lines of code form a huge perennial flow of self-reproducing writing. This is a new phase of history characterized by the extreme complexity of the systems in which we live. It is a global paradigm shift in technology, society, economy, and politics.

If we do not organize complexity, complexity destroys us. For this reason, system engineering—the "daughter" of von Neumann, Wiener, and Forrester's cybernetics (see Printz 2018, figure 1.1)—plays a crucial role for the future of humanity. The main problem of system engineering is managing complexity and complex organizations. System engineering came out in the 1950s at the same time as the development of the first computers—it was the computer that led to the discovery of complexity and the need to organize it.[18] A software system is a complex system that has to manage many operations of different types at the same time and always in different situations (see the classic work by Brooks 1975). "Complexity" here concerns (a) the technical objects that we are able to build; (b) the projects that force many different actors to interact and collaborate; and (c) uses and users, that is, the semantics of technical objects. These last two levels are entirely human, non-technical levels, although they have a direct impact on the technical, i.e., on the first level.

Printz (2018, 126–36) distinguishes two major types of complexity in system engineering: (a) static complexity, which is related to the organization of the system, and (b) dynamic and behavioral complexity, which is related to the evolution of the system and its relationship with the environment. The study and analysis of complexity represent the basis of software engineering. I limit myself here to some bibliographical indications. For a history of cybernetics, computer science, and systems engineering from Wiener to von Neumann, I recommend these important books: Heims (1980), Dyson (2012), Aspray (1990), Printz (2018, chapters 1–3), Priestley (2011), and Haigh (2019). On the concept of complexity and the architecture of complexity, see the classic work by Simon ([1969] 1996, chapters 7–8).

A fundamental law of a systemic approach dictates that to interact with an environment, each system must have an abstract model of this environment that is independent of the physical and organic nature of the environment and the system (Printz 2018, 28). The Turing machine[19]—implemented by von Neumann's architecture—is the logical and mathematical metamodel of our informational systems. In other words, the Turing machine and von Neumann's architecture are automated self-reproducing metamodels that define the general grammar of informational systems, that is, what each system can do and how it can do it. The fundamental problem is that of translating the information that comes from the surrounding environment into a language that is adaptable to the logical and mathematical metamodel. In other words, the problem is not the Turing machine itself but understanding how to "talk" to this machine, how to give it "orders," that is, instructions for solving concrete problems. This requires building a system from the metamodel.

To better elaborate this point, I introduce the stack model (Figure 4.6), which is the result of the convergence of the Turing machine, von Neumann's architecture, and Shannon's theory of information. Each computer system is a stack of levels that interact with each other. Every computer, smartphone, or tablet works this way. Each level corresponds to a different type of complexity and, therefore, different properties, laws, and states that interact. This is the thesis of Anderson's famous article "More Is Different" (1972): at different levels of scale and complexity, new properties and rules emerge that are not deducible from the previous levels. The idea of emerging properties is the fundamental principle of system engineering: the architecture of a computer, smartphone, or tablet is a stack of different levels that are partly autonomous but interact and always produce new qualities and rules. This stack cannot be reduced to a few fixed laws. The only law is the breaking of symmetry. What is important is not the behavior of the individual levels but the behavior of the whole system. Each of the levels of the stack is a Turing machine. The entire stack is, therefore, a universal Turing machine embodied by the von Neumann architecture.

The Turing machine and von Neumann's ENIAC and ENDVAC[20] were radical novelties in the history of human technology and cannot be compared

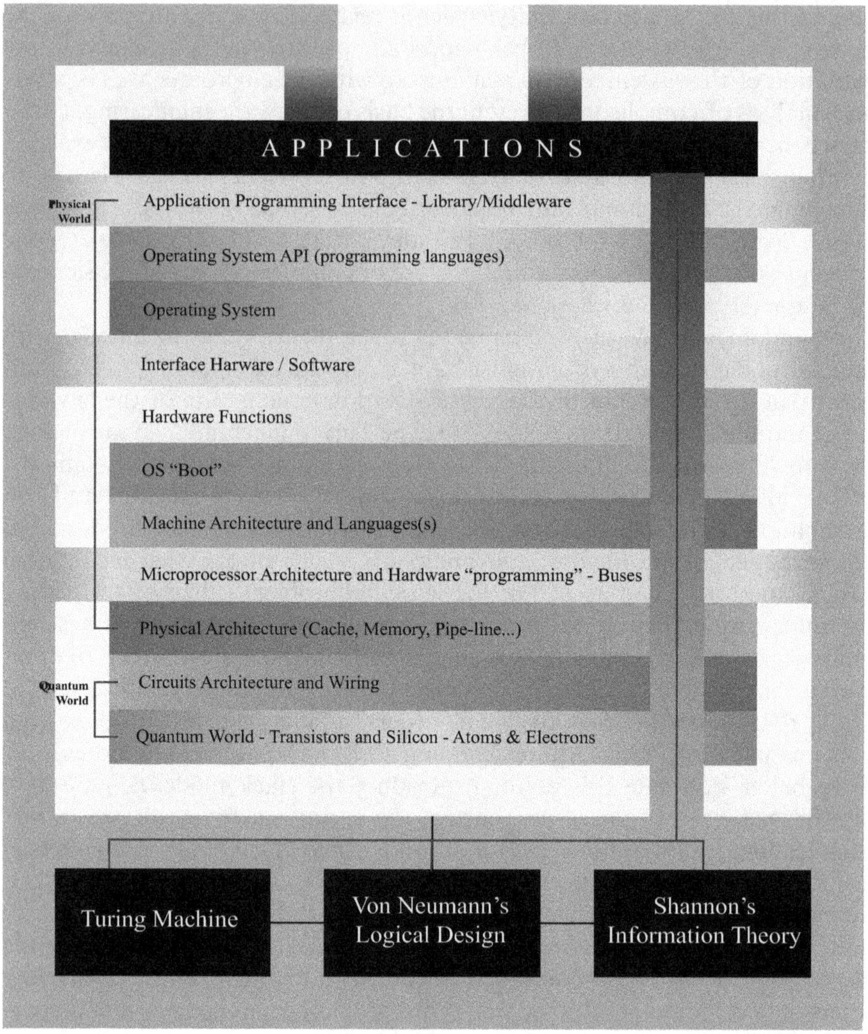

Figure 4.6 The stack. The basic structure of a software system.

to the theoretical and technical attempts of Babbage, Leibniz, or Pascal. Drawing this type of analogy indicates little understanding of the underlying logic and engineering issues. As Dyson (2012, 9) says, "the stored-program computer, as conceived by Alan Turing and John von Neumann, broke the distinction between numbers that *mean* things and numbers that *do* things." Turing and von Neumann invented a universal machine capable of adapting to any type of problem, thanks to the possibility of recording multiple programs in it. The stored program computer is a universal logical model independent of the physical dimension of the technical organs used by the machine. This model summarizes Russell and Frege's logic, Boole's logical calculation, as well as the Gödel results. The key idea is that "every method of systematic resolution, what we would call a process, can be performed by a universal Turing machine" (Printz 2018, 93; my translation). The ideas of Turing and von Neumann were the origin of a revolution that would lead to today's "cyberspace" (Printz 2018, 97–103; my translation).

In the stack, software itself—the programming language—appears only in the highest levels, which correspond to high levels of abstraction. But what is programming? Programming "has become a new way of thinking. [...] The act of programming, its 'performative' aspect, as linguists say, is at the heart of the systemic approach" (Printz 2018, 48; my translation). Software is not synonymous with algorithm. It is much more complex: it is "a new way of thinking." No algorithm exists and acts alone.

An algorithm always acts within a system and in connection with many other algorithms and with different types of languages and processes, which allows for interaction with databases and with the surrounding environment. This set of languages, operations, functions, control structures, surveillance structures, and structures for interaction with the environment (Printz 2006, 22–3) performs an essential task: the mediation between the human world, with its concrete requirements and problems (this set of constraints is called PESTEL), and the machines themselves, that is, the physical machinery. In other words, the fundamental objectives of software are (a) to develop a solution strategy for specific problems (for example, facial recognition), and (b) to translate this strategy into languages and structures that are understandable by the machine. In a software system, this translation process is realized by many different programs, languages, operations, and shared rules that define modules, applications, and groups of applications.[21] These applications work with different types of control systems, surveillance systems, and data structures (Printz 2006, 54–9). Indeed, the concepts of control, communication, and self-correction are essential in defining any system.

From a "textual" point of view, software can be considered an immense library organized into three main sectors. The first is that of books written in natural, informal language—this is the documentation part. The second is that of books written in high-level language, i.e., a language that is formal (Basic, Pascal, Python, Java, etc.) but can include some programmers' commentaries in natural language. These are the languages that serve the needs of programmers. There are also many more specific languages that concern, for example, memory, connection networks, and other devices. The third sector is that of the "texts" written by compilers, which are the translation of the books in the second sector into a language understandable and executable by machines. Each component of a software system is located in these three libraries and in each sector, but the content is described in different ways. The difficulty lies exactly here, or in the dialectic between a non-formal expression (expressed in natural language) and a formal expression (in a programming language). The crucial transition is that from the first to the third level of Figure 4.7, i.e., from the real problem to the software solution.

Software is essentially *problem solving* (see the classic work by Polya 1945). The software system must respond to end-user needs and goals (outer environment) and be adapted to the international criteria of software engineering (inner environment). Moreover, the activities of the groups of programmers must deal with political, economic, social, technical, ecological, and legal conditions (PESTEL). Programmers (obviously, according to the hierarchy in the company) study the problem, design a solution, and then define the system architecture. They follow all the usual technical management processes: decision analysis, technical planning, technical assessment, requirement management, risk management, configuration management, technical data management, interface management, etc. These activities produce a vast literature that is also organized into three parts, as we have seen above. This

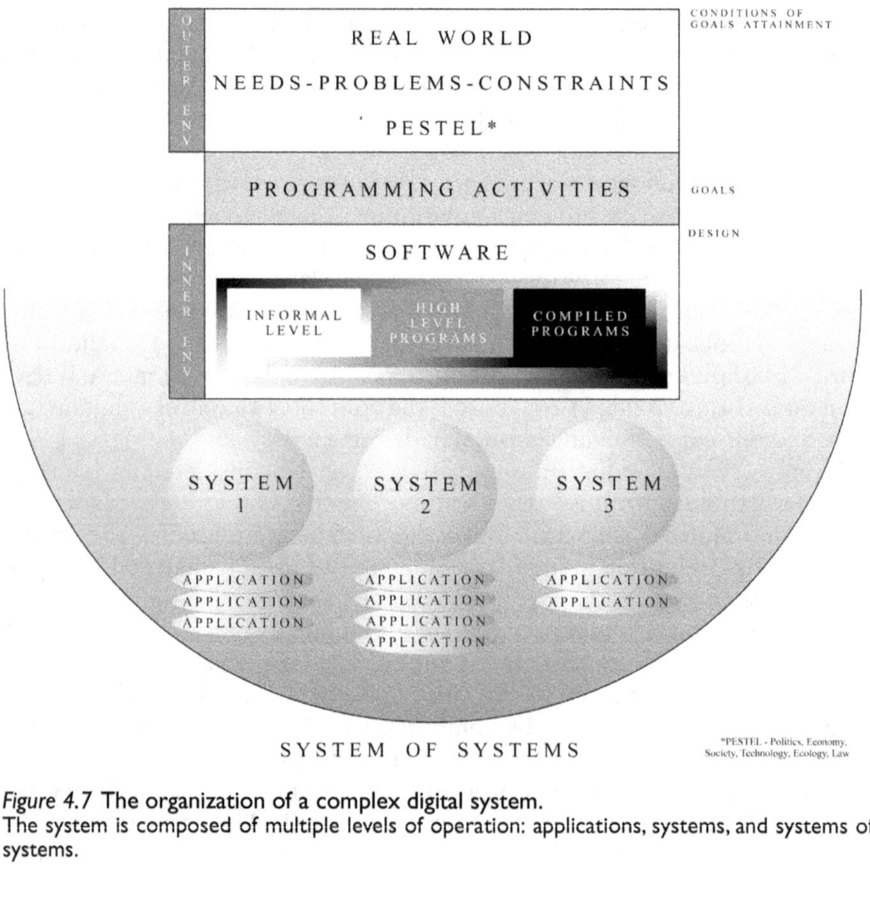

Figure 4.7 The organization of a complex digital system.
The system is composed of multiple levels of operation: applications, systems, and systems of systems.

literature is the content of which the system is comprised. Each part of the system (modules, applications, systems, and systems of systems) is comprised of pieces of this literature.

As Dyson (2012, 87, emphasis added) says, "von Neumann knew that the real challenge would be not building the computer, *but asking the right questions, in language intelligible to the machine.*" Does this mean that the programmer "speaks" to the machine? And what kind of language is software? From a linguistic point of view, software is *imperative*, that is, a set of commands, and *performative*, in the sense that "it does what it says," i.e., the statements translate into action simply by being formulated. Then, there is a third characteristic: software is *ego-less*, as Weinberg claims. In software, there is no ego that talks about itself and its life. The programmer neither talks about himself/herself nor tells a story. Its purpose is—in very simple terms—to translate a real problem and solution into a conceptual, logical, and linguistic structure that can be understood and executed by the machine.

An essential point must be underlined here. Software claims to be ego-less, *but it is not because it is and will always remain a human activity.* I am not saying that psychoanalysis can explain everything about software and AI. Obviously, it cannot. However, psychoanalysis can give us a new perspective on many problems, starting with the relationship between human beings and AI and what the pathological forms of this relationship are. This theoretical hypothesis is coherent and legitimate, as I have tried to demonstrate in the previous chapters. Furthermore, it contributes to an existing research field, namely, the psychology of programming (Weinberg 1971).

An engineer would undoubtedly say that we are saying crazy things. "Programming," the professional would say, "is a process well framed in rules and based solely on logic." However, because programming is also a human activity, it inevitably has a psychological dimension that plays a role on many levels. Programming is a social activity and, therefore, is affected by group psychological dynamics, as well as the personality of each programmer. "Programming is not just human behavior; it is complex human behavior" (Weinberg 1971, 40). This seems to me evident in the case I mentioned in the first chapter of this book, namely, the systems that manage university loans in the United States. With regard to emotions, Weinberg says: "Generally speaking, the computer is not interested in the emotional state of the programmer, although evidences of it may creep into his program" (54). In order to justify this claim, Weinberg points out that "a programmer would not really be a programmer who did not at some time consider his program as an esthetic object" (Weinberg 1971, 56). Humanists

> often contend that machines tend to dehumanize people by forcing them to have rigid personalities, but really, the contrary is true. [...] Relationships in which one party does all the giving or all the taking are

> not fully human and tend to produce personality distortions in one or the other.
>
> (Weinberg 1971, 67)

This is an essential point, which confirms our analyses in the second chapter. In programming design,

> in making our adjustments to our particular programming language, we can easily become attached to it simply because we now have so much invested in it. [...] We see this effect when we try to teach a programmer his second language.
>
> (Weinberg 1971, 68)

In order to understand these psychological dynamics, reading the code is not enough, says Weinberg; "we shall have to do more than read programs—we shall have to observe programs being made" (1971, 71). However, even observation might not be enough. Programming is such a complex activity that it cannot be understood using just one method. "One reason for these complexities is that the programmer himself seldom knows why he does what he does—which is a general problem in the study of any human behavior" (Weinberg 1971, 72).

Weinberg's analyses are important and useful; however, in my opinion, they do not get to the point. They appear rather as a series of considerations on certain aspects of the programmer's work. There is no systematic method, no basic theoretical structure. One question is crucial from my point of view: Can we hypothesize that the method of selecting programmers in large companies (see Ensmenger 2010, 59–75) has had a significant impact on the general personality of the programmer, e.g., the stereotype of the antisocial person obsessed with computers and mathematics? Can we say that this combination (selection method + stereotype) has greatly influenced the way our software is built and works?

Psychoanalysis can give us an anthropological framework within which to answer this question. Like any human behavior, programming also has unconscious dimensions that affect the functioning of the machine. Furthermore, as programming is a new form of human behavior, it also presents new unconscious dimensions never studied before. In this book, I claim that the concept of projective identification is an important theoretical tool for studying groups of programmers and their work in AI, in the manner of group psychoanalysis. There is a need for a new capacity of interpretation in order to analyze and interpret the dynamics of projective identification in groups of engineers and designers in AI and—beginning with this first work—the unconscious *habitus* that govern their work.

Notes

1 Simondon (2005) poses the philosophical problem of what the individual is. The two main ways in which the Western philosophical tradition has thought of the individual (atomist monism vs. Aristotelian hylomorphism) have both made the same mistake: they explained the thing with the thing itself, thus giving an absolute privilege to the concept of an already formed and constituted individual, without trying to problematize that being. For Simondon, however, the individual as such can never be the starting point. Instead, it is the result of a complex process starting from a pre-individual stage of being.

2 The statement is from an interview with Fortune.

3 www.forbes.com/sites/cognitiveworld/2020/02/07/biased-algorithms/.

4 vox.com/science-and-health/2019/1/23/18194717/alexandria-ocasio-cortez-ai-bias.

5 propublica.org/article/machine-bias-risk-assessments-in-criminal-sentencing.

6 science.sciencemag.org/content/356/6334/183.full.

7 technologyreview.com/2020/08/14/1006780/ai-gpt-3-fake-blog-reached-top-of-hacker-news/.

8 forbes.com/sites/tomvanderark/2020/08/17/welcome-to-human-computer-co-creation-what-gpt-3-means-for-education/#2618ed712490.

9 forbes.com/sites/tomvanderark/2020/08/17/welcome-to-human-computer-co-creation-what-gpt-3-means-for-education/#2618ed712490.

10 vox.com/science-and-health/2019/1/23/18194717/alexandria-ocasio-cortez-ai-bias.

11 wired.com/story/the-real-reason-tech-struggles-with-algorithmic-bias/.

12 wired.com/story/the-real-reason-tech-struggles-with-algorithmic-bias/.

13 wired.com/story/the-real-reason-tech-struggles-with-algorithmic-bias/.

14 brookings.edu/research/algorithmic-bias-detection-and-mitigation-best-practices-and-policies-to-reduce-consumer-harms/.

15 lanl.gov/discover/news-release-archive/2020/June/0608-artificial-brains.pdf.

16 lanl.gov/discover/news-release-archive/2020/June/0608-artificial-brains.pdf.

17 lanl.gov/discover/news-release-archive/2020/June/0608-artificial-brains.pdf.

18 The "complexity" emerges in the field of sciences with the works of Maxwell and Boltzmann on the kinetic theory of gas, whose aim is to describe the macroscopic behavior of gases starting from the atomic particles that constitute them.

19 I found what I consider the best definition of the Turing machine in Printz (2018, 147):

> The Turing machine is a mathematical artifact, necessary and sufficient for Turing's demonstration of the non-decidability of the decision problem formulated by D. Hilbert [...] but, in any case, it is not a computer model because for this, we would have to wait for von Neumann and his logical model.
>
> (my translation)

20 Also, the relevance of Gödel's results cannot be overestimated:

> In demonstrating that the property of being the code of a proof in PM [*Principia Mathematica*] is expressible inside PM, Gödel had to deal with many of the same issues that those designing programming languages and those writing programs in those languages would be facing.
>
> (Dyson 2012, 106)

21 Modules and applications are the basic cells of the system. An application is a set of modules assembled statically or dynamically. "An application is said to be integrated when all of its modules have satisfied all the integration criteria: an integrated application is an assembly of integrates. A non-integrated application has no use for the system user" (Printz 2006, 66; my translation).

References

Anderson, P. W. 1972. "More Is Different." *Science* 4047, no. 177: 393–6.

Aspray, W. 1990. *John von Neumann and the Origins of Modern Computing*. Cambridge, MA: MIT Press.

Barthélémy, J.-H. 2005. *Penser l'individuation. Simondon et la philosophie de la nature*. Paris: L'Harmattan.

Bertin, J. 1967. *Sémiologie graphique*. Paris/La Haye: Gauthier-Villars.

———. 1981. *Graphics and Graphic Information Processing*. Berlin: de Gruyter.

———. 1983. *Semiology of Graphics.* Madison: University of Wisconsin Press.

———. 2001. "Matrix Theory of Graphics." *Information Design Journal* 10, no. 1: 5–19.

Bourdieu, P. (avec la collaboration de Passeron, J.-C.). 1970. *La reproduction*. Paris: De Minuit.

Bourdieu, P. 1998. *Méditations pascaliennes*. Paris: Seuil.

———. 2000. *Esquisse d'une théorie de la pratique*. Paris: Seuil.

Brooks, F. 1975. *The Mythical Man-Month: Essays on Software Engineering*. Boston, MA: Addison-Wesley.

Castoriadis, C. 1975. *L'institution imaginaire de la société*. Paris: Seuil.

Coeckelbergh, M. 2020. *AI Ethics*. Cambridge, MA: MIT Press.

De Haan, Y., S. Kruikemeier, S. Lecheler, G. Smit, and R., Van der Nat. 2018. "When Does an Infographic Say More Than a Thousand Words? Audience Evaluations of News Visualizations," *Journalism Studies* 19, no. 9: 1293–312.

Dyson, G. 2012. *Turing's Cathedral. The Origins of the Digital Universe*. London: Penguin Books.

Elliott, E. 2012. *Reinvention*. London/New York: Routledge.

———. (2013) 2020. *Concepts of the Self*. London/New York: Routledge.

———. 2016. *Identity Troubles. An Introduction*. London/New York: Routledge.

———. 2018. *The Culture of AI: Everyday Life and the Digital Revolution*. London/New York: Routledge.

Elliott, E., and S. Frosh (eds). 1995. *Psychoanalysis in Contexts. Paths between Theory and Modern Culture*. London/New York: Routledge.

Ensmenger, N. 2010. *The Computer Boys Take Over*. Cambridge, MA: MIT Press.

Finn, E. 2017. *What Algorithms Want. Imagination in the Age of Computing*. Cambridge, MA: MIT Press.

Floridi, L. 2010. *Information. A Very Short Introduction*. Oxford: Oxford University Press.

———. 2011. *The Philosophy of Information*. Oxford: Oxford University Press.

———. 2016. *The Fourth Revolution. How the Infosphere is Reshaping Human Reality*. Oxford: Oxford University Press.

Floridi, L., N. Fresco, and G. Primiero. 2015. "On Malfunctioning Software." *Synthèse* 192, no. 4: 1199–220.

Fresco, N., and G. Primiero. 2013. "Miscomputation." *Philosophy and Technology* 26, no. 3: 253–72.

Freud, S. (1899) 1997. *The Interpretation of Dreams.* Reprint, Hertfordshire: Wordsworth.

———. 2018. *Totem and Taboo*. London: Dover Publications.

Friendly, M. 2008. "A Brief History of Data Visualization." In *Handbook of Data Visualization*, edited by C. Chen, W. Härdle, and A. Unwin. Berlin: Springer.

Fry, H. 2018. *Hello World. How to Be Human in the Age of the Machine.* London: Penguin Random House.

Goddard, G., B. Halligan, and P. Hegarty. 2012. *Reverberations. The Philosophy, Aesthetics and Politics of Noise*. London: Continuum.

Gray, J., L. Chambers, and L. Bounegru. 2012. *The Data Journalism Handbook*. Sebastopol: O'Reilly Media, Inc.

Grosman, J., and T. Reigeluth. 2019. "Perspectives on Algorithmic Normativities: Engineers, Objects, Activities." *Big Data & Society* 1: 1–6.

Haigh, T., ed. 2019. *Exploring the Early Digital*. Berlin: Springer.

Hainge, G. 2012. *Noise Matters: Towards an Ontology of Noise*. London: Continuum Publishing Corporation.

Hartmann, H. 1956. "The Development of the Ego Concept in Freud's Work." *The International Journal of Psychoanalysis* 37: 425–37.

Heims, S. J. 1980. *John von Neumann and Norbert Weiner from Mathematics to the Technologies of Life and Death*. Cambridge, MA: MIT Press.

Héran, F. 1987. "La seconde nature de l'habitus. Tradition philosophique et sens commun dans le langage sociologique." *Revue française de sociologie* 28, no. 3: 385–416.

Hermida, A., and M. Young. 2019. *Data Journalism and e Regeneration of the News*. Boca Raton, FL: CRC Press.

Holsanova, J., N. Holmberg, and K. Holmqvist. 2009. "Reading Information Graphics: The Role of Spatial Contiguity and Dual Attentional Guidance." *Applied Cognitive Psychology* 23, no. 9: 1215–26.

Hui, Y. 2019. *Recursivity and Contingency*. New York: Rowman & Littlefield.

Latour, B., and S. Woolgar. 1979. *Laboratory Life. The Construction of Scientific Facts*. Thousand Oaks, CA: Sage Publications.

Lima, M. 2014. *The Book of Trees. Visualizing Branches of Knowledge.* New York: Princeton Architectural Press.

Malaspina, C. 2018. *An Epistemology of Noise*. London: Bloomsbury Publishing.

Manovich, L. 2008. "Introduction to the Info-Aesthetics." http://manovich.net/content/04-projects/060-introduction-to-info-aesthetics/57-article-2008.pdf

Meirelles, I. 2013. *Design for Information*. Rockport: Rockport Custom Publishing.

Moses, R. 1987. "Projection, Identification, and Projective Identification: Their Relation to Political Process." In *Projection, Identification, Projective Identification*, edited by J. Sandler, 58–67. London: Routledge.

Munzner, T. 2014. *Visualization Analysis and Design*. Boca Raton, FL: CRC Press.

Neurath, M., and R. Kinross. 2009. *The Transformer: Principles of Making Isotype Charts*. London: Hyphen Press.

O'Neil, C. 2016. *Weapons of Math Destruction*. Washington: Crown Books.

Piccinini, G. 2018. *Physical Computation*. Oxford: Oxford University Press.

Polya, G. 1945. *How to Solve It*. Princeton: Princeton University Press.

Priestley, M. 2011. *A Science of Operations. Machines, Logic and the Invention of Programming*. Berlin: Springer.
Printz, J. 2006. *Architecture logicielle*. Paris: Dunod.
———. 2018. *Survivrons-nous à la technologie?* Paris: Les acteurs du savoir.
Primiero, G. 2020. *On the Foundations of Computing*. Oxford: Oxford University Press.
Riche, N., C. Hurter, N. Diakopoulos, and C. Sheelagh. 2018. *Data-Driven Storytelling*. Boca Raton, FL: CRC Press.
Ricoeur, P. 1983–1985. *Temps et récit*, vols. 1–3. Paris: Seuil.
Ritvo, L. B. 1965. "Darwin as the Source of Freud's Neo-Lamarckianism." *Journal of the American Psychoanalytic Association* 13: 499–517.
———. 1974. "The Impact of Darwin on Freud." *The Psychoanalytic Quarterly* 43: 177–92.
Rodighiero, D., and L. Cellard. 2019. "Self-Recognition in Data Visualization." *Espace-Temps* 8, no. 8.
Rodighiero, D., and A. Romele. 2020. "The Hermeneutic Circle of Data Visualization: The Case Study of the Affinity Map." https://pdfs.semanticsch-lar.org/1dc3/8398d0705c3350a70a975abbb5087d6b7051.pdf?_ga=2.37549158.753349493.1569967606-1517796743.1566339002
Romele, A. 2019. *Digital Hermeneutics. Philosophical Investigations in New Media and Technologies*. London/New York: Routledge.
Shannon, C., and W. Weaver. 1964. *The Mathematical Theory of Communication*. Urbana: University of Illinois Press.
Simon, H. (1969) 1996. *The Science of the Artificial*. Cambridge, MA: MIT Press.
Simondon, G. 2005. *L'individuation à la lumière des notions de forme et d'information*. Paris: Millon.
———. 2007. *L'individuation psychique et collective*. Paris: Aubier.
———. 2010. *Communication et information*. Paris: Editions de la Transparence.
———. 2012. *Du mode d'existence des objets techniques*. Paris: Aubier.
Sloterdijk, P. 2013. *You Must Change Your Life*. Cambridge, MA: Polity Press.
Solms, M. 1997. "What Is Consciousness?" *Journal of the American Psychoanalytic Association* 45, no. 3: 681–703.
Tufte, E. 1990. *Envisioning Information*. Cheshire: Graphics Press.
———. 2001. *The Visual Display of Quantitative Information*. Cheshire: Graphic Press.
Vial, S. 2013. *L'être et l'écran*. Paris: Puf.
Ware, C. 2012. *Information Visualization: Perception for Design*. Amsterdam: Elsevier.
Watkins, Y., Kim, E., Sornborger, A., and G. T. Kenyon. 2020. "Using Sinusoidally-Modulated Noise as a Surrogate for Slow-Wave Sleep to Accomplish Stable Unsupervised Dictionary Learning in a Spike-Based Sparse Coding Model." Working paper, Computer Vision Foundation. https://openaccess.thecvf.com/content_CVPRW_2020/papers/Watkins_Using_Sinusoidally-Modulated_Noise_as_a_Surrogate_for_Slow-Wave_Sleep_to_CVPRW_2020_paper.pdf
Weinberg, G. 1971. *The Psychology of Computer Programming*. New York: Van Nostrand Reinhold Company.
Wiener, N. 1961. *Cybernetics or Control and Communication in the Animal and the Machine*. New York: The MIT Press.
Wilkins, I. Forthcoming. *Irreversible Noise*. Falmouth: Urbanomic.
Zuboff, S. 2018. *The Age of the Surveillance Capitalism*. London: Profile Books.

Chapter 5

A Freudian computer

Neuropsychoanalysis and affective neuroscience as a framework to understand artificial general intelligence

5.1 Introduction

The thesis of this chapter is that neuropsychoanalysis and affective neuroscience can provide a new paradigm for AI and, in particular, for artificial general intelligence (AGI). Here, I refer to the works of Mark Solms and Jaak Panksepp in particular. Neuropsychoanalysis and affective neuroscience give us a precise answer to the enigma of mind/brain dualism by highlighting the constant interaction of these two dimensions. I will try to show how this new approach is key to conceptualizing AGI. I will try to use very simple language to avoid unnecessary complications and ensure that my explanation is understandable by all. Of course, the initiation of an interdisciplinary dialogue between such different fields of study (AI and neuropsychoanalysis) requires maximum humility, mutual respect, and attention in order to avoid hasty solutions. My argument is that AGI is possible and recent developments in neuromorphic engineering and computational biology are going in that direction. For this reason, a reference framework is needed.

The structure of the chapter is as follows. In the first part (Sections 5.2 and 5.3), I provide a definition of the fundamental aspects of neuropsychoanalysis and its relationships with affective neuroscience. In the second part (Sections 5.4 and 5.5), I illustrate the neuropsychoanalytic model of the mind/brain relationship. In particular, I analyze Panksepp's theory of the basic affective states. In the third part (Section 5.6), I define some basic design principles for an AGI model based on neuropsychoanalysis and affective neuroscience.

Why do we need a new approach to AGI? I will answer this question with two remarks. The first is this: one essential point that AI research must consider is that a method based merely on the physical imitation of the brain is wrong. There are two reasons for this: the first is that our knowledge of the brain is still very limited; the second is that even assuming that we could properly reconstruct each cell of our brain and its functions, something would still be missing, namely, the mind. Therefore, we need a model that can hold these two dimensions together, the mind and the brain. The imitation of anatomical mechanisms and the psychological expression of these mechanisms must go hand in hand.

The second remark is as follows: so far, research in AI has mainly focused on the activities of the cerebral cortex and the main cognitive functions (language, logic, memory, cognition, etc.). Now, the time has come to conceptualize and develop an AGI of the subcortical. Neuropsychoanalysis and affective neuroscience affirm the centrality of emotions and affect—in the subcortical area of the brain where primary processes (instincts, emotions, feelings) are located—in psychic and cognitive activity. My hypothesis is that an AGI system inspired by neuropsychoanalysis and affective neuroscience must be based on the modeling and simulating of the seven basic affective systems analyzed by Panksepp. Panksepp's work—which is also very important for neuropsychoanalysis—establishes a theoretical framework to use in the development of this hypothesis.

An important clarification should be made. Today, there is a lot of talk about affective computing. The relationship between emotion and AI is an exceptionally vast research field founded by the important and controversial book by Rosalind Picard, *Affective Computing* (1997). What does it mean for a computer to "have emotions" or feelings? In general, when we talk about affective computing, we mean three connected things: the ways in which (a) a computational system can recognize and simulate human emotions; (b) a computational system can respond to human emotions; and (c) a computational system can express emotions spontaneously or use them in a positive way in the decision-making process (see El-Nasr, Yen, and Ioerger 2000; Erol et al. 2019; Fogel and Kvedar 2018; Schuller and Schuller 2018; Shibata, Yoshida, and Yamato 1997). "More specifically, affective computing involves the recognition, interpretation, replication, and potentially the manipulation of human emotions by computers and social robots" (Yonck 2017, 5). Experts agree that artificial emotional intelligence is a continuously developing research field and that it will have a decisive importance in economies and societies of the future. However, artificial emotional intelligence will also pose new ethical and legal problems—and also new dangers, such as psychological manipulation (see Picard 1997, chapter 4). The study of biomimetics and hybrid systems (biological and technological) that analyze the possibility of building robots capable of reproducing the versatility of the human organism (see Prescott and Lepora 2018) is also connected to this immense research field.

How is my research different from the affective computing approach? This chapter does not intend to provide an overview of the affective computing debate. To accomplish such a task would require not a few paragraphs but an entire book. However, I will develop some critical considerations regarding Picard's concept of emotions and feelings. In my opinion, Picard still remains too tied to a cognitivist conception of mind, and this prevents her from considering emotion as such. Following Panksepp (1998) and Panksepp and Biven (2012), I argue that emotion is an intrinsic function of the brain, not the reflection or the derivative of the higher cognitive functions. There exist basic instinctual systems that are phylogenetic memories and evolutionary tools for living. If we do not fully understand these systems, we cannot understand the brain/mind relationship. Human emotionality has an intelligence, a structure; it is not simply the mechanical answers to a series of random situations. A subcortical AI should be capable not only of reproducing the basic human affective systems but also of building the most elaborate cortical functions, such as learning or language, from the previous ones.

From this point of view, an AGI system able to "imitate" or "simulate" these basic affective systems must be thought of in a way radically different from classical methods. It must be based primarily on mathematically modeling the behavior of neurotransmitters, which act in different ways in the basic affective systems—for example, dopamine in the system that Panksepp calls "seeking."

5.2 A case study: Anella

I take as the starting point of my analysis the case of the neural network system Anella, inspired by Freudian metapsychology. In the next section, I will try to expand the line traced by Anella.

The core idea of Anella is that thought is the result of the self-organization of a set of words, i.e., connected speech elements (Jorion [1989] 2012, 15). Each linguistic act must be understood as a sequential path in a space of words, i.e., a set of elements of a lexicon.[1] Talking means going through a lexicon. The method chosen to allow the AI system to use and follow the lexicon is associative. Inspired by the concept of "induced association" by Jung and "free association" by Freud ([1899] 1997), the creators of Anella argue that there are connections between the elements of a lexicon that are not rigid rules but "privileged passages" or "free connections" (Jorion [1989] 2012, 41; my translation). The connections link the speech elements, understood as signifiers, i.e., they are considered in their materiality and sensitivity. Language is a set of connections of signifiers. Semantics comes from the development of these connections (79–80).

Let us imagine a lexicon. We store its elements in the system memory. Different types of connections are automatically created between these speech elements, i.e., the signifiers. Simply run the system, and the connections can develop independently. The creators of Anella represent the connections by means of a graph in which the arrows represent the signifiers and the vertices the connections (Jorion [1989] 2012, 97–102). But is this enough? No, it cannot be. The connections are not all the same. The set of automatic connections is only part of the memory system. Other connections, made by the development of the system (and therefore not automatic but due to learning), have to be added. Furthermore, automatic connections can be very different, as in natural language.

How can we differentiate connections and paths in the Anella network? We attribute an energetic quantum to each connection between signifiers. Older connections will have more energy, while newer connections will have less. The older connections correspond to the most archaic memories of the system, to which the most primitive and strongest drives are connected—what the creators of Anella call "germs of belief" (Jorion [1989] 2012, 103; my translation). The more the connections grow, the more the energy available in the system decreases. Newer connections involve less energy investment and, therefore, create less imbalance in the system. In this way, the system self-regulates; its goal is to preserve the equilibrium. This means that the system makes differentiations: the connections of the highest energy quantum are made inaccessible, while those of the lowest are accessible and work. This is the mechanism of repression in Freud's terms. The connections of the highest energy quantum are not canceled; they remain in the system, but they are isolated, including the memory cells that they connect. For this reason, the system must create new connections that are increasingly complex and tortuous. I do not want to complicate things further by introducing the Freudian distinction between primary removal and secondary removal or the Lacanian concept of *forclusion* (Lacan 1981, 558; Jorion [1989] 2012, 117–8). The idea is clear: the self-regulation of the system creates two different zones in the system memory. If we follow the analogy with Freud, we will say that there are three zones: the unconscious (inaccessible connections), the conscious (accessible connections), and the preconscious (the part of the conscious that remains in the potential state in a certain session of system activity).

Let us take a closer look at the affective dynamics. Two forces are associated with each connection. The first is its energetic quantum, i.e., the drive connected to it. The second is the internal regulation of the system: in each activity carried out by the system, connection with the high-energy quantum is associated with an energy quantum of the opposite sign, which is capable of reducing the voltage and, therefore, of maintaining the energy balance of the system.

The meaning of a term in Anella is the constellation of associative connections in which the term is located, which is the same as the dictionary definition. "The delimitation of a term—what the dictionary proposes as its meaning—corresponds to the delimitation of a subnet of the memory network" (Jorion [1989] 2012, 132; my translation). A phrase, however, corresponds to a completely new constellation of connections that cannot be reduced to the sum of the meanings of the terms within it. The meaning of a phrase corresponds to the compatibility between its constellation and others in the network. In other words, the meaning of a phrase consists in its compatibility with the evolution of the system, i.e., with the previous and subsequent sets of phrases. Different constellations correspond to different affective dynamics. As a result, semantics, logic, and the theory of truth are emerging phenomena, i.e., *a posteriori* formalizations of the memory network

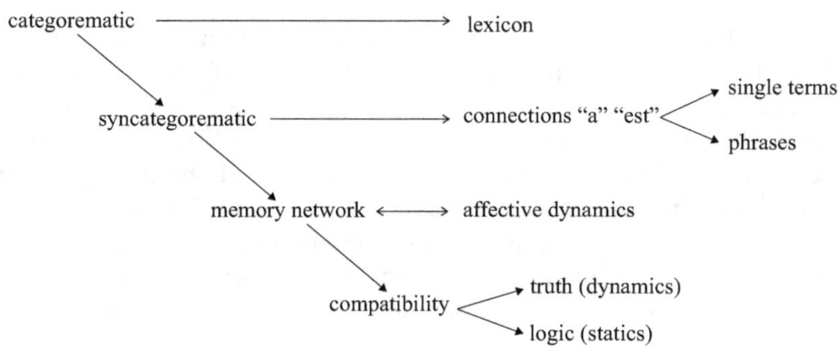

Figure 5.1 The structure of Anella.
Each linguistic act is represented as a sequential path in a space of words, i.e., a set of elements of a lexicon.

and of the associations that make it up. They are phenomena that spontaneously emerge from the connections in the memory network.

We can summarize this progression in the scheme in Figure 5.1.

The learning concerns the system in action. For Anella, learning means discovering, through interaction with the surrounding environment, new terms that are not present in its network. Once a new term is discovered, Anella looks for ways to integrate it into the network and then connects it with other terms. Learning consists in creating new connections. A term is properly learned once it is inserted into a constellation of terms and receives a sense and emotional dynamic. In learning, the system is reconfigured or extended both linguistically and affectively. The newer parts of the system are constantly evolving, while the older ones cannot be modified so easily. The system grows and self-modifies. The order in which the terms are learned—and, therefore, the series of changes that the memory network undergoes over time—determines how the system evolves. Thus, the order in which the terms are learned represents the history of the system and its "personality" (Jorion [1989] 2012, 151). An AI system is historically constructed, just as a human being is. "In general, terms are given to us: learning essentially consists in reproducing in our memory the associative links conveyed by the culture that surrounds us" (154; my translation). In human beings, this evolutionary process is much more complex because it includes images, smells, sounds, and tactile perceptions: all these elements connect to the memory network and modify it.

The Anella model is still limited. It concerns only language and, therefore, a very limited area of human cognitive activities. Furthermore, it does not offer any real analysis of the emotions or the relations between them and the brain. Can neuropsychoanalysis allow us to improve the line drawn by Anella? Can the emotional neurosciences allow us to improve the Freudian point of view on emotion?

5.3 Neuropsychoanalysis: beyond Freud, with Freud

The main advocates of the neuropsychoanalytic point of view (Solms, Kaplan-Solms, and Turnbull) argue that their point of view on the mind/brain question is the same as Freud's and call this approach "dual-aspect monism." Their thesis is that the mind is a unique reality. Nonetheless, we cannot directly access it. To describe and understand the mind, we must draw inferences (to build models) based on two limited forms of experience: first-person subjective experience (psychology) and third-person study of brain structures and functions (neuroscience). These two forms of experience are independent and have the same value, but neither is able to explain this unique reality, which can be called mind/brain, in a complete way. If we look at it with our eyes (or perceive it using other sensory organs), we are presented with a brain, a biological organ like many others. However, if we look at it with the eyes of our subjective consciousness, we come into contact with mental states such as sadness, desire, pleasure, etc. It is, therefore, necessary to keep both points of view (subjective and objective) open and build a dynamic parallelism between them. We will never find a thought, a memory, or an emotion in a piece of brain tissue; we will find brain cells and nothing else. Meanings and intentionality are not reducible to neurons. According to dual-aspect monism, the mind can be distinguished from the brain only from the perceptual perspective. If we admit the existence of a single entity X "behind" the terms "mind" and "brain," then we can say that (a) the mind is X perceived subjectively, that is, through my own consciousness, and (b) the brain is X perceived objectively, that is, through external perception and objectifying methods of sciences.

Neuropsychoanalysis tries to connect the X-object to the X-subject. In this way, neuropsychoanalysis does not intend to reduce the mind to the brain; even if it has been accused of biologism, it does not intend to reduce everything to biochemical processes and anatomy. All mental phenomena require a biological correlate; this is indisputable. However, saying this does not mean entirely reducing mental phenomena and their meaning to supposed biological correlates. The biological and psychological dimensions must be kept together; they are two sources of information that must be considered to be of the same value. The biological dimension can inform us about mental phenomena and their meaning by supplying us with additional information. Psychoanalysis can return the favor by enriching research in neuroscience. Neuropsychoanalysis holds that psychoanalysis can intervene in the patient's brain and modify it, as did the traumatic experiences of his or her past. People with depressive symptoms, psychic trauma, obsessions, shyness, or compulsive and self-destructive impulses always have detectable neuronal anomalies that are strictly related to the functioning of their mind. The meanings are every bit as real as neurons or chemicals, and they can have effects in reality. For this reason, psychoanalysis, acting on the minds of patients, can also cure their neuronal anomalies. Conversely, studying people with brain injuries can help us better understand the development of the mind.

Neuropsychoanalysis does not set out to prove that Freud was always right. Instead, it claims to finish the work started by Freud. Indeed, Freud began his career as a neuroscientist and neurologist (Sulloway 1979, chapter 1). He had a specific and wide-ranging scientific program, but it was largely conditioned by the limits of the neuroscientific methods available at the time. For Freud, psychoanalysis is more than just a hermeneutics of mental life. The separation between psychoanalysis and neuroscience was, for him, only a pragmatic, strategic, and temporary solution; it was motivated by the lack of knowledge about the brain in its time. However, as Freud repeats in several passages, the inevitable progress of neuroscience would sooner or later lead to bridging the gap between the two disciplines and to providing an organic basis to the discoveries of psychoanalysis (Solms and Turnbull 2002). In other words, Freud was dissatisfied with the clinical–anatomical method of his time and, therefore, developed his analytical method independently of neuroscience from 1895 to 1939. He eagerly awaited the progress of neuroscience and biology, and, for this reason, he sought confrontation, dialogue, and cooperation with these sciences (Solms and Saling 1990).

From Freud's time to today, things have changed a lot. Today, we can verify the validity of Freud's basic statements through appropriate scientific observations. Modern knowledge and methods for studying the brain, which are much more developed than in Freud's time, allow us to improve and finish Freud's endeavor. In the past 20 years, neuroscience has not only experienced exponential growth but also changed in character, thanks to technological advances. In particular, the critique of the behavioristic (focused on the observable patterns) and cognitive (based on the thesis that the human mind is essentially information processing, as are perception and learning) models of mind has led to a broader vision that includes emotions and feelings, a vision which acknowledges the connection of the mind to a body that acts and perceives within a social and technological environment. Both the behavioristic and cognitive models undermine the importance of emotions and feelings.

Evidence of this turning point can be found in numerous works: Benedetti (2010), Damasio (1994), Decety and Ickes (2009), Gallese (2009), LeDoux (1996), and Panksepp (1998). Furthermore, Lurija's important work (1976) also demonstrated the possibility of renewing the psychoanalytic method through neuroscience. In particular, Solms (2000) and Kaplan and Solms (2000) underlined the importance of the Lurija method, which entails the abandonment of a rigid localization of cognitive functions in favor of a much more integrated approach to the mind. This is the so-called "dynamic localization method," according to which each of the complex mental activities (memory, imagination, thought, etc.) cannot be rigidly localized in a single area of the brain. On the contrary, during instances of these activities, many different areas of the brain activate each time and in different ways.

It should never be forgotten, however, that the debate on neuropsychoanalysis is broad and complex. Much research in neuroscience claims that the Freudian dream theory (but not only this) is completely wrong (Hobson 2007). Furthermore, according to many psychoanalysts (see Blass and Carmeli 2007; Edelson 1986; Pulver 2003), neuroscience is irrelevant to psychoanalysis, and this is because neuroscience has nothing to say about our mental meanings and their interpretation, which fall within the field of psychoanalysis. Knowing the biological basis of mental processes explains nothing about the meanings that make up our life; it would be like wanting to explain software on the basis of hardware. As I mentioned above, in the following, I will limit myself to drawing upon the works of two authors: Solms and Panksepp.

5.4 The neuropsychoanalytic model of the mind

Neuropsychoanalysis proposes a general model of how the human mental apparatus, as it is conceived by psychoanalysis, can be represented in brain tissues. It is a hypothetical model based on current knowledge of the brain and on limited empirical data. The theoretical points of reference for this operation are Lurija's work and Freudian metapsychology. Mental functions are not rigidly localized in individual areas of the brain, but instead distributed in several different areas. Each area contributes in its own way.

I depict the neuropsychoanalytic model of the mind in the scheme in Figure 5.2.

At the center of our scheme is the P-C system: perceptual consciousness. This system has two surfaces: an external one directed toward the world around the brain (it is divided into different areas of specialization: sight, hearing, kinesthesia, and tactile sensation) and an internal one directed toward the processes taking place inside the body. The first surface is located in the posterior part of the cortex (although numerous subcortical structures contribute to the processing of the stimulus). The second surface is connected to the limbic system and to a series of deeper, subcortical brain structures, which represent the oldest part of the brain. These are the only two sources of stimuli for the brain; these are external reality (R) and internal reality (ID). For neuropsychoanalysis, therefore, consciousness is nothing abstract or metaphysical. It is the set of our external and internal perceptions, that is, the connection between the data we have about the external world and the way in which these data modify us.

According to Solms (2008), the intermediate zone between internal and external perception corresponds to areas of the unimodal cortex that filter, record, and structure information by using connections, associations, and classifications. These areas are located in the posterior parts of the cortex. Associations and connections can be of different types, depending on the

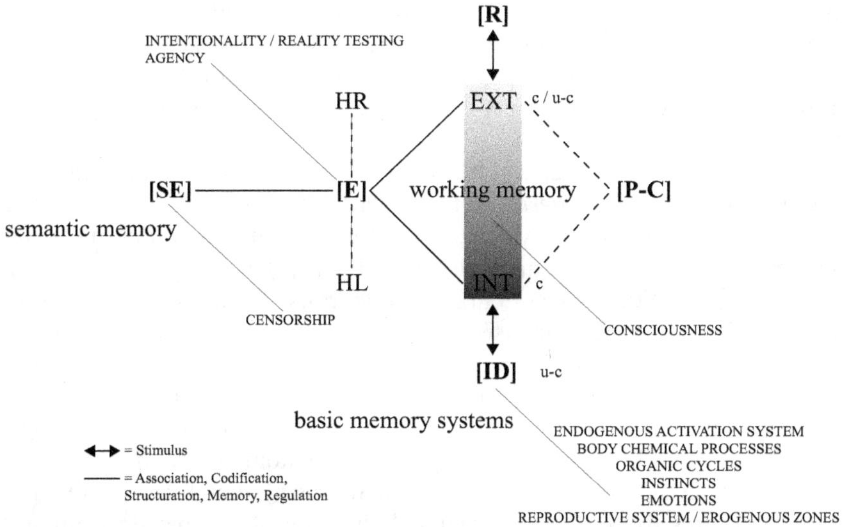

Figure 5.2 The neuropsychoanalytic model of mind.
E = ego; ID = Freudian id, which is mainly located in the subcortical areas of the brain; P-C = perceptive systems which have two surfaces, internal (INT) and external (EXT); SE = super-ego, which mainly corresponds to the prefrontal cortex of the brain. Other abbreviations: c = cortical tissue; u-c = subcortical tissue. HL = left hemisphere; HR = right hemisphere.

type of memory involved in the process: working memory, episodic memory, procedural memory, semantic memory, etc. The information is recorded in different ways. Through memory, the brain develops intentionality—the ability to plan its actions and, therefore, to act in the world. In addition, it develops the capacity for thought, that is, deferring action or satisfaction of need at a given moment.

Solms links this intermediate zone to the Freudian notion of ego. In Freudian metapsychology, in fact, the ego is that instance which must mediate between external and internal reality; it is precisely that part of the id that has been modified by external reality (natural and social) in the course of evolution. Freud also saw in the ego a series of memory structures through which experiences are connected and recorded. These connections are not predetermined; rather, they develop over time. The progressive stabilization of the connections gives rise to the primary cognitive functions, such as thought, language, logic, attention, calculation, and imagination. The right hemisphere organizes information by using spatial relationships, while the left hemisphere does so by following temporal relationships. Two important articles by Kandell (1979, 1983) explain the way in which these processes develop at the cellular level. The ego is a continuous dynamic connective process of which the constant evolution depends on numerous variables. The crucial function of the ego is to work as a barrier to stimuli. If there were no ego to filter and organize information, the human brain would be overwhelmed by stimuli; consequently, it would be in a state of perennial excitement.

Note that our diagram distinguishes a west and a south of the ego. The south is the id, whereas the west is the super-ego. The id is the origin of ego and the super-ego.

The id is our deep, visceral biological dimension, what is also called the "internal milieu" (*milieu intérieur*), and which includes different bodily systems, such as the musculoskeletal system, the immune system, the endocrine system, the chemical processes, organic cycles, etc. The way the brain perceives changes occurring in this biological system is what neuropsychoanalysis calls "internal perception"; this is the immense field of instincts, feelings, and emotions to which psychoanalysis ascribes a predominant role in psychic activity. Internal perception consists in the activation of deep and ancient brain structures (the limbic system and subcortical brain structures) connected to the biological dimension of the body and the mechanisms of adaptation. It is important to note that the neurons that make up the limbic system and the subcortical brain structures work very differently than those of the perceptive-mnestic systems of the cortex (Solms 1996). These neurons, which are both digital and analogue, respond not only to discrete stimuli but also to gradual state changes. I will say more on the affective systems in the next section.

To the west of the ego is the super-ego. According to Solms (1996), the super-ego can be connected to certain regions of the prefrontal lobe and precisely to those regions that connect the prefrontal part of the brain with the limbic system. These regions act as a filter that censors the needs of the instinctual pole of the mind. Their type of memory is called "semantic memory" and mainly concerns social conventions. In line with what Freud says, the super-ego arises from the internalization of behavior and value schemes in the social context. Here, I should underline one interesting point about the super-ego that concerns our path in this book: the super-ego and the projective identification process are very similar because both arise from a destructuring of the ego. Assuming that AI arises from a projective identification process, then it is a phenomenon that is deeply connected to a "crisis" of the ego and represents its coherent evolution "after" the super-ego. Thus, AI is part of mind/brain evolution.

Before concluding this section, one puzzle must be solved. The source of activation of internal perception is the id, that is, the vital biological systems that compose the human organism. Does this mean that the id is conscious and that we have perception of the id? Such a statement would have enormous consequences for the psychoanalytic investigation. Do we have perception of the unconscious? Is the unconscious conscious? Where does the unconscious fit into our scheme (Figure 5.2)? Solms claims that the id is the source of all forms of consciousness: "This constant 'presence' of feeling is the background *subject* of all cognition, without which consciousness of perception and cognition *could not exist*" (Solms 2013, 16). The point is that this basic form of emotional consciousness is not fully translated into the more complex systems of the ego and the super-ego; therefore, it remains invisible. This is the Freudian theme of the repressed: "If we retain Freud's view that repression concerns representational processes, it seems reasonable to suggest that repression must involve withdrawal of *declarative* consciousness" (Solms 2013, 17). In a nutshell, the id has no access to declarative consciousness.

5.5 The primitive affective states, or the basic human values

Let me begin with a passage from Damasio (1994, 158–9):

> It does not seem sensible to leave emotions and feelings out of any overall concept of mind. Yet respectable scientific accounts of cognition do precisely that, by failing to include emotions and feelings in their treatment of cognitive systems. [...] [E]motions and feelings are considered elusive entities, unfit to share the stage with the tangible contents of the thoughts they nonetheless qualify. This strict view, which excludes emotion from mainstream cognitive science, has a counterpart in the no less traditional brain-sciences view [...] that emotions and feelings arise in the brain's down-under, in as subcortical as a subcortical process can be, while the stuff that those emotions and feelings qualify arises in the neocortex. I cannot endorse these views. First, it is apparent that emotion is played out under the control of both subcortical and neocortical structures. Second, and perhaps more important, feelings are just as cognitive as any other perceptual image, and just as dependent on cerebral-cortex processing as any other image.

The organism is composed of a series of structures that stand in relation to each other. Homeostasis is the set of coordinated and partly automatic physiological, biological, and chemical processes that are indispensable for maintaining the stability of the state of the organism and, thus, guaranteeing survival. Examples include the regulation of temperature and heart rate, the concentration of oxygen in the blood, the structure of the musculature, skin tone, and metabolism. According to Damasio (1999), emotions are closely connected to homeostasis; they are biological phenomena produced by neuronal configurations in order to guarantee homeostasis. The brain influences and modifies the body by regulating it. The aim is adaptation and survival, that is, to create advantageous conditions for the organism in certain situations. For example, fear causes the acceleration of heart rate in situations of danger. An external situation (the danger) activates certain regions of the brain that produce, through the release of chemicals or neurotransmitters, a series of modifications of the body (the acceleration of the heartbeat, the movement of the legs, etc.). In response to the brain, the body changes its internal regulation mechanisms and adapts to the new situation to survive. Damasio distinguishes primary emotions (joy, sadness, fear, anger, surprise, and disgust) from secondary emotions, which are more complex. In addition, there are the background feelings, such as well-being, malaise, calm, and tension, expressed in the details of posture and in the way of moving the body. With the somatic marker hypothesis, Damasio has shown that emotions play a role of primary importance in cognitive processes (Damasio 1994, 45–9). Furthermore, consciousness itself is closely connected to emotion, feeling, and homeostasis; it is a more refined and effective form of realizing homeostasis in the face of the challenges posed by the surrounding environment (see Damasio 1999, chapter 10).

Based on the study of animals and the comparison between animals and humans, Panksepp (1998) offers us a much more elaborate and complete theory of emotions than does Damasio. According to Panksepp, Damasio is still a victim of the cognitivist prejudice because he continues to think that emotions are a variant of higher cognitive processes, i.e., the result of a sort of "rereading" of them by the cortex. Such cognitive prejudice can also be found in Rolls (1999, 2005), where there are no basic affective states; emotions are the products of cognitive activity—for example, the ability to verbalize or conceptualize assessments is considered a necessary condition for emotional experience.

For Panksepp, these theories are full of problems and contradictions. For example, how can a cognitive state give rise to an affective experience? Panksepp, who is closer to neuropsychoanalysis than is Damasio, has identified the existence of an ancestral core of emotional processes that underlie any form of psychic activity, whether unconscious or conscious. Panksepp argues that emotions are intrinsic functions of the subcortical brain, a feature that humans have in common with animals. Emotions, or affects, are "ancient brain processes for encoding value—heuristics of the brain for making snap judgments as to what will enhance or detract from survival" (Panksepp and Biven 2012, 31–2). These basic affective systems are not cognitive at all; they "are made up of neuroanatomies and neurochemistries that are remarkably similar across all mammalian species" (Panksepp and Biven 2012, 4).

In general, Panksepp distinguishes three levels of brain activity:

(a) The primary process, which includes the most basic affects;
(b) The secondary process, such as learning and behavioral and evolutionary habits; and
(c) The tertiary process, which includes executive cognitive functions (thoughts and planning).

The primary process activities are organized into three areas: emotional affects, homeostatic affects, and sensory affects. The homeostatic affects concern internal biological cycles (the need to eat or defecate, for example) that allow homeostasis. Sensory affects are reactions to sensations experienced from the outside; they are exteroceptive sensory-triggered feelings of pleasure and displeasure/disgust. Emotional affects (also called "emotion action systems" or "intentions-in-actions") are the oldest and most complex. Panksepp organizes these affects into seven systems: seeking, fear, rage, lust, care, panic/grief, and play. These systems are described by Panksepp as real physical circuits present in the most ancient and deep parts of the brain, the subcortical area, which activate certain reactions and behaviors (for example, the rat escapes the smell of predators, and this pushes it to look for another area in which to feed) and, therefore, forms of learning. They are instinctive (automatic reactions) and evolutionary (the result of a long natural selection process). Panksepp argues that raw affects are the fundamental basis of any brain activity; the mind is essentially emotional, and raw affects tend to shape any other cognitive activity. "Most prominently, it looks like the seeking urge may be recruited by the other emotional systems. It is required for everything the animal does; one could conceptualize it in psychoanalytic terms as the main source of libidinal energy" (Panksepp 2008, 165).

In other terms, raw affects are ancient brain processes for coding values, that is, heuristic operations of the brain used to make rapid assessments of what, in real situations, increases or decreases the chances of survival. It is possible to give a description of these brain circuits (see Panksepp and Biven 2012, 75–6). They can interact with and be influenced by cognitive states in very complex ways, but they do not presuppose them. They are, to use a Pankseppian expression, "a flexible guide for living" (Panksepp and Biven 2012, 43).

The crucial point of Panksepp's approach is that basic emotions have no cognitive content; therefore, they cannot be understood from a cognitive-computational point of view. Instead, they must be dealt with on their own specific terms.

> The widespread claim that affects are just a variant of cognitions seems little more than a word game to me, even though I certainly accept that the many valenced (good and bad) feelings of the nervous system are always interacting with cognitions (imagination, learning, memory, thoughts) within the full complexities of most human and animal minds.
> (Panksepp and Biven 2012, 489)

Therefore, Panksepp claims that cognitions are often "handmaidens," or emissaries, of the affects, not the opposite. Cognition is

> from the neocortex, which is the brain's outermost layer and the part that is evolutionarily newest. This indicates that the capacity for affective experience evolved long before the complex cognitive abilities that allow

> animals to navigate complex environmental situations. It is also noteworthy that the deeper evolutionary location of the affective systems within the brain renders them less vulnerable to injury, which may also highlight the fact that they are more ancient survival functions than are the cognitive systems.
>
> (Panksepp and Biven 2012, 43–4)

The affective brain is activated by subcortical regions; it is less computational and more analogue. Affects are automatic, instinctual, and innate processes, but individual behavior, education, and culture cannot change them. The cognitive brain is instead more neocortical, more computational and digital, and responds to different chemicals than the affective brain does. The will, imagination, language, logic, and the most advanced forms of consciousness are connected to the cognitive dimension and all it entails (on the distinction of "two brains," see Kahneman 2011). We cannot understand secondary and tertiary functions if we do not understand primary functions first. This is also confirmed by other data: subcortical neurons function very differently from those of the regions of the neocortex (see Panksepp and Biven 2012, 50).

This distinction between the two brains is important, and it is the reason that leads me to criticize Picard's point of view. Like Damasio, Picard remains too tied to a cognitive conception of emotions and affects. According to Picard, emotions are generated by cognitive activity (a thought, the knowledge of a state of things, etc.) (see Picard 1997, 65–6). With this, in my view, Picard does not grasp the essence of human emotional life. Rather, this point of view implies that emotion is not something intrinsic to the human brain but something built from cognitive reflections operated by a human or a machine. On the contrary, Panksepp says that emotion is intrinsic to the brain and that it is the brain that produces the physiological reactions of the body. Emotion—the dimension of raw affects and primary processes—has its own specific intelligence, which cannot be reduced to digits and computation, that is, to the tertiary process.

The second crucial aspect that emerges from Panksepp's research is the complexity of emotions and the brain. We cannot reduce emotions and affects to simple bipolar systems based on the pairs of opposites charge/discharge and pleasure/displeasure. They are much more complex; they cannot be reduced to the on/off mechanics of neurons. "The simple-minded neuron–doctrine view of brain function, which is currently the easiest brain model to apply in AI/robotics, under-represents what biological brains really do" (Panksepp 2008, 163). Each basic affective system acts in a different way according to very complex chemical and neurochemical dynamics and equilibria that we do not yet fully grasp. This puts one of the central acquisitions of psychoanalysis and neuropsychoanalysis into question. Emotions cannot be explained in a dualistic way according to a series of oppositions arranged on three levels: energetic, perceptive, and motor. Each of the fundamental affective systems generates positive or negative states; however, in reality, the distinction between pleasure and displeasure is not clear-cut. There are many intermediate or even superimposed states (such that the same situation generates pleasure in one case and, in another, displeasure).

Therefore, the crucial condition for an AGI based on the neuropsychoanalysis model is the design of a computational system capable of simulating the seven basic affective systems analyzed by Panksepp. This system should consist of multiple logic systems intertwined with each other—Matte Blanco (1975) can give important suggestions about this. The mechanisms of human raw affects must be the fundamental basis of AGI. For this reason, research on animals can be a key element.

5.6 A Freudian computer: sketches

In their attempt to build an AI system based on neuropsychoanalysis, Dietrich, Ulieru, and Kastner (2007) consider the modeling of emotions as an open problem: "The scientific literature of neuropsychoanalysts does not provide an answer clear enough for the kind of model engineers need for technical realization" (4). I suggest that Panksepp's topography of emotions can solve this problem and give us a clear indication of what an emotion is and how basic affective systems work. A vision of AI based too much on brain simulation risks remaining too tied to a sterile form of behaviorism. The study of the subcortical brain regions supplies us with a different framework. It should be noted that the purpose of this chapter is not to make technical proposals but to indicate some design principles for an AGI based on neuropsychoanalysis and affective neuroscience.

Shanahan (2015) distinguishes five general characteristics that an AGI system must possess:

- Embodiment;
- The ability to interact with a dynamic environment;
- The ability to learn;
- Common sense, or the ability to understand the principles of the everyday world, in particular, the social environment; and
- Creativity, or the ability to generate new behavior or invent new things.

In the light of the neuropsychoanalytic model of the mind, how can we interpret these characteristics? I propose the following reformulation:

- Embodiment;
- The ability to have a relationship with the body and, therefore, "affective depth," i.e., the production of raw affective states (Panksepp's model);

The ego, which is the ability to learn, must mediate between external (natural and social) stimuli and internal (the basic affective states) stimuli;
The super-ego, on the other hand, which is the ability to understand essential social principles, develops from the ego and performs a censorship function (this is the root of language);
Creativity is the ability to change behavior in order to meet internal needs (basic affective states and other feelings) despite the objections of the super-ego and the demands of the ego; and
The stratification of different types of memories in the different distinct systems (affective, ego, super-ego); the memory traces are continuously translated from one system to another.

Hence, an AGI system based on neuropsychoanalysis and affective neuroscience needs

To create computational models capable of simulating basic affective systems, for which teams made up of engineers, psychoanalysts, and neuroscientists are required;
To develop research on animals, which is the primary source of Panksepp's work; and
To develop new forms of materials for AI.

5.6.1 The foundations of AGI

At the root of the AGI system, seven systems are necessary in order to simulate the seven basic affective systems in mammals. "In order to simulate the operations of the human mind, we must consider both the genetic and epigenetic construction of the human brain. We must be clear about what is genetically fundamental and what is epigenetically derivative" (Panksepp 2008, 149). Perhaps

> the most accurate simulations need to get the genetically-provided subsystems properly represented in Read only Memory (ROM) as key operating systems and to configure Random Access Memory Space (RAM) in such a way that developmental epigenetic programming can simulate the natural ecologically and culturally constrained developmental landscapes that transpire in neocortical maturation of the human child.
>
> (149)

Each basic affective system could be described "in terms of 'state-spaces' that regulate 'information-processing' algorithms" (149).

Can such affective-emotional properties of biological brains be emulated by machines? Panksepp is skeptical: "Only future work can tell" (Panksepp

2008, 149). For Panksepp, simulating the ancient visceral nervous system is problematic:

> a deep understanding of the subcortical tools for living and learning [is] the biggest challenge for any credible future simulation of the fuller complexities of mind. The cognitive aspects may be comparatively easy challenges since many follow the rules of propositional logic.
>
> (150)

The crucial question is whether and how an algorithm can simulate the complex subcortical circuits. What kind of logic should we follow? This issue "may require a complete rethinking of where we need to begin to construct the ground floor of mind" (152). Panksepp has "no confidence that the natural reality of those processes can be computed with any existing procedures that artificial intelligence has offered for our consideration" (152).

Hypothetical computational systems that mimic the subcortical brain must mostly reproduce analogue, not digital, processes-or digital-analogue processes. This means that they must have a general regulatory function; they must describe the equilibrium of the whole system at a given time. This remains a challenge.

5.6.2 A system composed of multiple systems

Let us assume that we can model the seven basic emotional systems. Signals from outside are conveyed to these seven systems, which process them. An interface collects, synthesizes, and stores the outputs in order to then produce what I call an "affective image" of the system. This is the first level of learning and memorization.

The affective image is then translated into the higher system, which collects the fundamental cognitive functions, those that we can identify with the ego: feelings, vision, language, planning, attention, etc. Each of these networks learns from external data but does so from the orientation of the affective image. For example, the affective image regulates assimilated words and their connection, or the level of attention and the type of planning. Furthermore, the higher system must also be able to develop more complex feelings; in this, Picard's work can be useful. The logical models used for this system should be increasingly complex: from classical logic to paraconsistent systems. The types of memory used should also be very different. The outputs of these higher systems are unified and recorded by an interface that I call the "*ego image*".

The third level of our AGI system corresponds to the super-ego. This part of the system is the most complex because it has to do with social conventions and language. Therefore, it must know how to use both of the previous images in order to (a) distinguish social and non-social agents in a given context; (b) grasp the patterns of behavior of social agents; and (c) regulate its behavior in relation to these patterns or conventions. It acts, essentially, as a censor in the relationship between the emotional image and the ego image. The super-ego is also formed by a series of systems corresponding to the different contexts experienced by the system; they represent, therefore, a pure learning process. The work of the different systems is conveyed in what I call the "*super-ego image*".

The idea of a multiplicity of systems made up of other systems, all different but connected, is essential to my model. This idea takes up a fundamental intuition of Freud in the second part of the seventh chapter of *The Interpretation of Dreams*: the psychic apparatus is not a monolith but a set of many different systems. The memory traces are continually written, rewritten, and associated in different systems (conscious, preconscious, unconscious) in different ways—Freud speaks of "diversified fixation." The repressed is what cannot be translated in the passage from one system to another. As I noted above, Solms (1996) distinguishes several memory systems.

Following this idea, I posit that the unconscious of the system is not only the organism and the emotional system but also the repressed in the translation process from one system to another. Something is lost; it is something that is not translated, cannot be translated, or is translated badly. I claim that this something "lost" becomes noise. Therefore, it has a dual existence: on the one hand, it continues to "live" its normal existence in a system; on the other, it now also "lives" in the new system but in another form, that of noise that "disturbs" the system. A regression from information to noise takes place. Untranslated or badly translated information turns into noise. I also hypothesize the formation of "noise nuclei," namely, large quantities of that which is repressed, which deform an entire area of the system activity.

We can, therefore, represent our system as a set of different types of memories through which data are written and rewritten several times (Figure 5.3). When these translation processes stop or are hindered, information becomes noise. What hinders or stops translation processes? Algorithmic biases, unconscious human processes, or inconsistency between systems.

On the neuropsychoanalytic level, nothing prevents us from saying that our AGI system (Figure 5.4) can develop a proto-self, a nuclear consciousness, an extended consciousness, and even an unconscious in a classically Freudian sense. In other words, nothing prevents us from saying that this system has a cerebral and psychic experience that has a high degree of correlation with our cerebral and psychic experience.

5.6.3 A body for AGI

Damasio seems inclined to believe that we can create artifacts capable of reproducing emotions and feelings, namely, the formal mechanisms of consciousness at the neurological level (1999, 377). However, he also believes

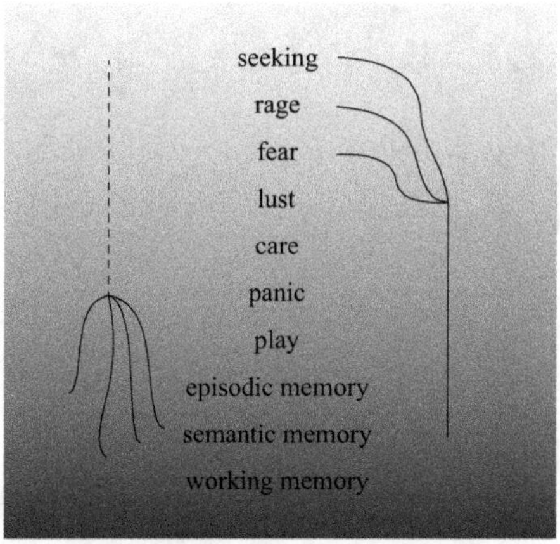

Figure 5.3 The system can be represented as a set of memory systems.
The same data are written and rewritten in each of these systems. The dotted line represents the translations of data from the subcortical systems to the higher cortical systems. The curved lines represent the noise; the translation process stops in a certain system and produces noise. The black line represents the translations of data from cortical to subcortical systems. As before, curved lines represent the noise.

that these artifacts would only be able to simulate the "appearances" of the emotion, that is, to reproduce some automatisms but not human feelings. The problem, Damasio claims, is that emotions and feelings are deeply connected to the body, to the flesh, something that silicon cannot reproduce. I think that this idea is very similar to Dreyfus' position. The American philosopher affirms that computers, which have neither a body, a childhood, nor a culture, cannot acquire intelligence in the proper sense. Dreyfus (1972) argues that much of human knowledge is tacit; therefore, it cannot be articulated into a program. The project of strong AI, that is, AI similar to the human being, is, therefore, impossible. This argument has been further developed by Fjelland (2020), who claims that causal knowledge is an important part of human-like intelligence, and computers cannot handle causality because they cannot intervene in the world.

On this point, however, more advanced research could really change things. I am not talking about biorobotics or biomedical engineering research but about the creation and development of the first biological robots, that is, robots made of programmable biological matter. In this regard, Kriegman et al. (2019) open a completely new path. They present the results of research that led to the creation and development of Xenobots, that is, the first tiny robots made entirely of biological tissues. Xenobots are a new lifeform on our planet. Researchers used an evolutionary algorithm to simulate the design of robots. They then selected the best models. These models have been tested to

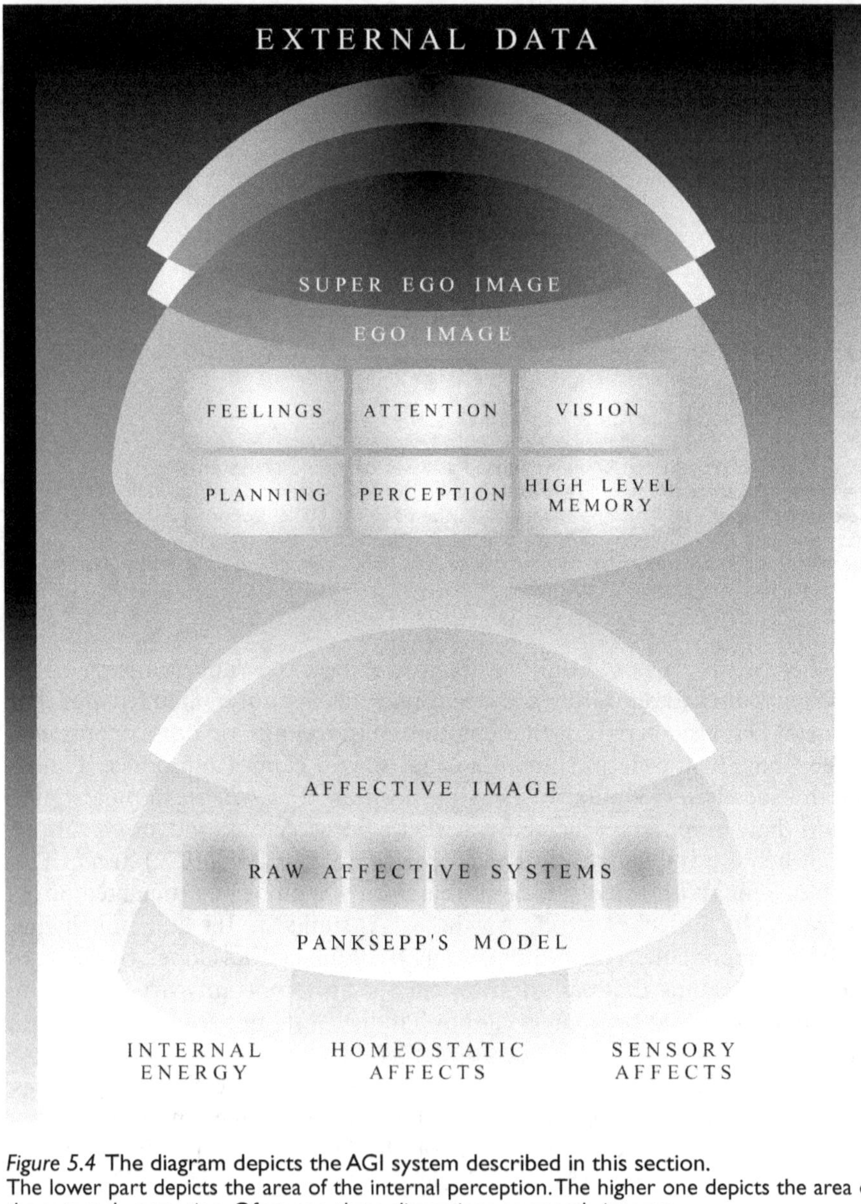

Figure 5.4 The diagram depicts the AGI system described in this section.
The lower part depicts the area of the internal perception. The higher one depicts the area of the external perception. Of course, these dimensions constantly interact.

make them increasingly capable of adapting to real situations; the researchers subjected them to large quantities of noise in order to understand whether, in normal situations, they would be able to maintain an intended behavior or not. The transition from the design to the implementation phase took place through the use of embryonic stem cells of *Xenopus laevis*, a type of frog. The cells were assembled and developed by a computer and then programmed to perform certain functions. "Programmed" means that the cells were assembled into a finite series of configurations corresponding to certain movements and functions in an aqueous environment. These microorganisms are neither animals nor traditional robots. They have a heart and skin. If they are damaged, they can repair themselves and survive for at least 10 days. They are assembled by the computer and programmed to behave according to models. The Xenobot is an organism in all respects, except that it is based on an artificial design. As it has a body, it has *bodily senses*, and, consequently, homeostatic and sensory affects. This solves many of the problems associated with robot embodiment (see Dietrich et al. 2008, 150).

The principle of biological programming could be used to build new computational systems and program the seven basic affective systems theorized by Panksepp into cells (or groups of cells). Powerful learning and evolutionary algorithms would allow us to understand how these cells evolve and whether they develop feelings and thoughts like those of humans or other animals.

5.7 Conclusions

I consider the theses developed in this chapter to be the beginning of a research program on the possibility of AGI based on the simulation of the subcortical areas of the brain. Only future investigations will establish the merits and drawbacks of the ideas developed here. The central theoretical hypothesis of this chapter is that AGI is possible only if the main cognitive functions are based on computational systems capable of adequately simulating the raw affective systems that humans share with other mammals. AGI must not be based on the imitation of the behavior of humans but on the simulation of their seven basic affective systems. This means that the design of AGI must overcome the boundaries between technology and biology, between silicon and flesh. My conclusion is that the classic argument of Dreyfus (1972; Dreyfus and Dreyfus 1986) is wrong: AI is already in our world, shapes our culture, and forces us to challenge our humanity. Dreyfus' argument is based on a too-rigid cognitive vision of the human being and mind.

Panksepp's work offers us a philosophical revolution: emotions and instincts are not "ghosts" of the mind that cannot be studied scientifically because they are too subjective and changeable. Rather, they are ancestral values that guide us in our life and make consciousness possible. Damasio (2010) also now approaches this point of view with an open mind after having supported more cognitivist positions. The raw affective systems are the preverbal and pre-intentional DNA of every human activity; thanks to neuroscience, they can be studied scientifically. They represent an archaic intelligence, autonomous "wild thought." AI must enter this revolution.

Note

1 By "elements of a lexicon," Jorion ([1989] 2012, 163) essentially means proper nouns, common nouns, adjectives, and verbs, i.e., *categorematic* terms. This represents all that is used in association processes. However, terms such as "but," "or," "and," "therefore," "nor," etc., which are called *syncategorematic* terms, as well as verbs that do not function as copulae or auxiliaries, are not included in association processes. Anella reduces all types of logical connections to two forms of connections ("est" and "a"; I follow the French formulation), and this allows it to break down the definition of a name into a series of associative links belonging to one or the other form. The whole functioning of Anella depends on the reduction of the *syncategorematic* terms to their basic association forms "est" and "a," as described in Jorion (73–91).

References

Benedetti, F. 2010. *The Patient's Brain*. Oxford: Oxford University Press.

Blass, R. B., and Z. Carmeli. 2007. "The Case against Neuropsychoanalysis: On Fallacies Underlying Psychoanalysis' Latest Scientific Trend and Its Negative Impact on Psychoanalytic Discourse." *International Journal of Psychoanalysis* 88, no. 1: 19–40.

Damasio, A. 1994. *Descartes' Error*. New York: Putnam.

———. 1999. *The Strange Order of Things*. New York: Pantheon.

———. 2010. *Self Comes to Mind. Constructing the Conscious Brain*. New York: Random House.

Decety, J., and W. J. Ickes. 2009. *The Social Neuroscience of Empathy*. Cambridge, MA: MIT Press.

Dietrich, D., M. Ulieru, and W. Kastner. 2007. "Considering a Technical Realization of a Neuro-psychoanalytical Model of the Mind. A Theoretical Framework." In *2007 5th IEEE International Conference on Industrial Informatics*. Vienna, Austria.

Dietrich, D., G. Fodor, G. Zucker, and D. Bruckner, eds. 2008. *Simulating the Mind: A Technical Neuropsychoanalytical Approach*. Berlin: Springer.

Dreyfus, H. L. 1972. *What Computers Can't Do*. New York: Harper & Row.

Dreyfus, H. L., and S. E. Dreyfus. 1986. *Mind Over Machine*. Oxford: Basil Blackwell

Edelson, M. 1986. "The Convergence of Psychoanalysis and Neuroscience: Illusion and Reality." *Contemporary Psychoanalysis* 22, no. 4: 479–519.

El-Nasr, M. S., J. Yen, and T. R. Ioerger. 2000. "FLAME: Fuzzy Logic Adaptive Model of Emotions." *Autonomous Agent and Multi-Agents Systems* 3: 219–57.

Erol, B., A. Majumdar, P. Benavidez, P. Rad, K. R. Choo, and M. Jamshidi. 2019. "Toward Artificial Emotional Intelligence for Cooperative Social Human–Machine Interaction." *IEEE Transactions on Computational Social Systems* 7, no. 1: 234–46.

Fjelland, R. 2020. "Why General Artificial Intelligence Will Not Be Realized." *Humanities and Social Sciences Communications* 7: 1–9.

Fogel, A., and J. Kvedar. 2018. "Artificial Intelligence Powers Digital Medicine." *Digital Medicine* 1, no. 5.

Freud, S. (1899) 1997. *The Interpretation of Dreams.* Reprint, Hertfordshire: Wordsworth.

Gallese, V. 2009. "The Two Sides of Mimesis: Girard's Mimetic Theory, Embodied Simulation and Social Identification." *Journal of Consciousness Studies* 16, no. 4: 21–44.

Hobson, J. A. 2007. "Wake Up or Dream On? Six Questions for Turnbull and Solms." *Cortex* 43: 1113–5.

Jorion, P. (1989) 2012. *Principies des systems intelligentes.* Broissieux Bellecombe-en-Bauges: Ed. Croquant.

Kandell, E. R. 1979. "Psychotherapy and the Single Synapse." *The New England Journal of Medicine* 301, no. 19: 1028–37.

———. 1983. "From Metapsychology to Molecular Biology: Explorations into the Nature of Anxiety." *American Journal of Psychiatry* 140, no. 10: 1277–93.

Kahneman, D. 2011. *Thinking Fast and Slow.* New York: Penguin Books.

Kaplan, K., and Solms, M. 2000. *Clinical Studies in Neuro-Psychanalysis.* Madison, CT: International Universities Press.

Kriegman, S., D. Blackiston, M. Levin, and J. Bongard. 2019. "A Scalable Pipeline for Designing Reconfigurable Organisms." *Proceedings of the National Academy of Sciences* 117, no. 4: 1853–9.

Lacan, J. 1981. *Le séminaire: les psychoses 1955–56. Livre 3.* Paris: Seuil.

LeDoux, J. 1996. *The Emotional Brain.* New York: Simon & Schuster.

Lurija, A. R. 1976. *The Working Brain.* New York: Basic Books.

Matte Blanco, I. 1975. *The Unconscious as Infinite Sets. An Essay in Bi-Logic.* London: Gerald Duckworth & Company Ltd.

Panksepp, J. 1998. *Affective Neuroscience. The Foundations of Human and Animal Emotions.* Oxford: Oxford University Press.

———. 2008. "Simulating the Primal Affective Mentalities of the Mammalian Brain: A Fugue on the Emotional Feelings of Mental Life and Implications for AI-Robotics." In *Simulating the Mind: A Technical Neuropsychoanalytical Approach*, edited by D. Dietrich, G. Fodor, G. Zucker, and D. Bruckner. Berlin: Springer.

Panksepp, J., and Biven, L. 2012. *The Archeology of Mind. Neuroevolutionary Origins of Human Emotions.* New York: W. W. Norton.

Picard, R. 1997. *Affective Computing.* Cambridge, MA: MIT Press.

Prescott, T. J., and N. Lepora, 2018. *Living Machines. A Handbook of Research in Biomimetics and Biohybrid Systems.* Oxford: Oxford University Press.

Pulver, S. E. 2003. "On the Astonishing Clinical Irrelevance of Neuroscience." *Journal of the American Psychoanalytic Association* 51, no. 3: 755–72.

Rolls, E. T. 1999. *The Brain and Emotion.* Oxford: Oxford University Press.

———. 2005. *Emotion Explained.* Oxford: Oxford University Press.

Shanahan, M. 2015. *The Technological Singularity.* Cambridge, MA: MIT Press.

Shibata, T., M. Yoshida, and J. Yamato. 1997. "Artificial Emotional Creature for Human-Machine Interaction." In *1997 IEEE International Conference on Systems, Man, and Cybernetics. Computational Cybernetics and Simulation*, 2269–74. Orlando, FL, USA.

Schuller, D., and B. W. Schuller. 2018. "The Age of Artificial Emotional Intelligence." *Computer* 51, no. 9: 38–46.

Solms, M. 1996, "Towards an Anatomy of the Unconscious." *Journal of Clinical Psychoanalysis* 5, no. 3: 331–67.
———. 2000. "Freud, Luria and the Clinical Method." *Psychoanalytic History* 2: 76–109.
———. 2008. "Repression: A Neuropsychoanalytic Hypothesis." www.veoh.com/watch/v6319112tnjW7EJH
———. 2013. "The Unconscious Id." *Neuropsychoanalysis* 15, no. 1: 5–19.
Solms, M., and M. Saling, eds. 1990. *A Moment of Transition. Two Neuroscientific Articles by Sigmund Freud.* London: The Institute of Psychoanalysis.
Solms, M., and O. Turnbull. 2002. *The Brain and the Inner World. An Introduction to the Neuroscience of Subjective Experience*. London: Karnac Books.
Sulloway, F. 1979. *Freud. Biologist of the Mind.* Cambridge, MA: Harvard University Press.
Yonck, R. 2017. *Hearth of the Machine. Our Future in a World of Artificial Emotional Intelligence*. New York: Arcade.

Chapter 6

Conclusions

Toward an ethnographic and psychoanalytic study of AI

We are living an unprecedented revolution. The Anthropocene, the geological age characterized by the impact of human activities on nature, which began with the invention and spread of the steam engine, is about to end. According to Lovelock (2019), Newcomen's steam engine, which gave rise to the Industrial Revolution, represented the first occasion on which solar energy was directly converted into mechanical energy. The steam engine then spread due to its effectiveness and profitability, thanks to a natural selection process that works for machines in the same way as for organisms. This process gave rise to a previously unthinkable transformation of the world. Today, we are facing an ever-accelerating and more connected world made up of huge city-states in which solar energy is transformed into information—the extraordinary book by Khanna (2016) gives perhaps the most complete view of this phenomenon. "Cities are the most visible signs of the power of the Anthropocene to transform our planet. Night photographs of the Earth taken from satellites show brilliant dots, strings and flashes of light clustered together" (Lovelock 2019, 53). The most recent development of the Anthropocene, that is, the century from the end of the American Civil War in 1870 to 1970, has produced the most impressive transformation of human life ever: an unrepeatable "golden century" (Gordon 2016).

Why should we be at the end of the Anthropocene? Lovelock responds with this very clear thesis: Earth is an old planet and, like all old ones, it is fragile. "Now, an asteroid impact or a volcano could destroy much of the organic life the Earth now carries" (Lovelock 2019, 59). Earth is a planet that is getting warmer, and this excessive warming is a threat to any kind of life. "So 47°C sets the limit for any kind of life on an ocean planet like the Earth. Once this temperature is passed, even silicon-based intelligence would face an impossible environment" (Lovelock 2019, 64). The regulation of temperature is essential for life, and this is Gaia's main task—according to Lovelock's hypothesis (2016). However, precisely for this reason, intelligence becomes essential; it is only thanks to an intelligent use of resources that Gaia is able to preserve itself and survive. The transition from humans to cyborgs is, therefore, inevitable; the more Gaia ages, the more the intelligence needed to keep her alive in the face of any "astronomical crisis" increases. The cyborgs imagined by Lovelock are machines capable of programming themselves; therefore, they are entirely free from human control. The human being has triggered a process: we just have to wait to see the new cyborg culture arise. It will be useless to try to communicate with these machines because they will have such fast minds that we will not be able to understand them—perhaps they will communicate with us in some way. "I expect their form of communication will be telepathic" (Lovelock 2019, 79). This will be the "Novacene."

Lovelock's hypotheses about the Novacene are often close to science fiction. Yet, they follow a strong logic, that of Gaia. Gaia's survival and homeostasis determine everything; humans and cyborgs cannot fail to adapt to it if they hope to continue to exist. This aspect positively characterizes Lovelock's hypotheses: there is no fear of a "singularity" in the form of a dramatic event. Increasing intelligence for survival is a necessity for Gaia; humans will be replaced by cyborgs without upheavals or conflicts à la *The Matrix*. Electronic life will take over without destroying humans because human life will continue to serve to ensure Gaia's self-regulation. Regardless of their scientific value, Lovelock's hypotheses demonstrate the importance of the AI theme for the future of humanity. An ethics of AI is not enough. We need a global approach to AI that is not reduced to merely its control and limitation. AI is not simply technology: it is culture, namely, a means at the disposal of the human being to construct its own identity and way of living in the universe. Will AI be able to save a capitalism in crisis? Will AI be able to save humanity from a future like the one envisaged by *Interstellar*, in which humans do not fight for money or prestige, but for food?

The goal of this book was not to establish a doctrine; rather, it was to develop a new AI research project. More than defining territories, the book aimed to open new paths to research and experimentation. The algorithmic unconscious is a concept that we can use to explain machine behavior. The validity of a theoretical hypothesis is also demonstrated by its fruitfulness, i.e., the opportunities it offers. The theoretical framework described in this book not only gives us good reasons to consider the hypothesis of an algorithmic unconscious as legitimate and necessary, but it also describes an interesting new approach to AI. This new approach can be defined as a "therapy." It is possible to develop psychoanalytic techniques that allow us to study the projective identification processes that are activated and transmitted through AI. I think that this kind of analysis could be usefully integrated into data analytics practices.

A good AI "therapist" studies what I called "the dynamics of projective identification." This means that the therapist has to work with groups of professional programmers and designers in order to understand the internal dynamics of these groups and the types of projective identifications within them. The organization of the groups and the work within each of them must be developed throughout the software construction process in each of its phases. Conflicts, pressures, and unconscious tensions in the group, the set of unconscious motivations, and the ability of team members to self-observe and self-understand can affect technical work. Projective identification is a concept that can help us gain a complete and unified view of all the aspects that are intrapsychic and interpersonal (Ogden 1982, 114–6). The study of group psychoanalysis from a theoretical, technical, and clinical point of view is, therefore, an essential tool (Romano 2017).

The analyst's first task is to clarify. The concept of projective identification is a framework for organizing and dynamically formulating the complex interplay between humans and machines. In human groups of programmers and designers, there are some projective identifications that have nothing to do with the machine, while others directly affect the design process and, therefore, influence the analyst's work. The analyst must isolate the latter projective identifications and provide a precise description. The analysis of the transference and countertransference is the core of this part of the work. However, the connections between projective identification and transference or countertransference dynamics are very complex and are beyond the scope of this work (see Ogden 1982, 68–73).

The work with human groups is only a first step in AI therapy. The organization of hybrid groups composed of humans and AI is equally important. Observing the dynamics—through interviews, tests, and free dialogues, depending on the type of AI considered—is fundamental to understanding the different types of projective identifications that develop from humans to machines, from machines to humans, and from machines to machines. It is essential to organize increasingly complex groups. Building an appropriate setting to observe and analyze AI interactions and behavior is another crucial aspect of this part of the work. One possible approach is to build and use special AI systems to understand these dynamics. Interpretation work is another very complex and difficult aspect to define without the appropriate groundwork. The AI analyst must first understand whether the interpretation of projective identifications should be verbal or non-verbal, whether it should be revealed to machines or humans, or both, and at what times.

Another essential task of the AI analyst is that of "reading" software. The AI analyst must understand the program and relate its structure and characteristics to the results of the work with the groups. Reading a program is not like reading a book, of course. It is a more conceptual reading, which must be organized around the aforementioned seven "hermeneutic areas." The analyst must relate the study of these "hermeneutic areas" to the dynamics of projective identification analyzed in the work with the groups. These two dimensions must go hand in hand. The analyst should refer also to the critical code studies (see Marino 2020).

However, the analyst's activity has to go beyond meetings with groups and interpretations. The analyst must also be an ethnographer and able to observe work processes. Being among computer scientists and technicians and observing them during their daily work is fundamental. I would say that it is a prerequisite of the entire analysis as it enables the analyst to truly understand what AI is, as well as what the relationship between technicians and AI is. In this phase, the observation and analysis of bugs, miscomputations, and algorithmic bias provide useful material for assessment when working with groups.

Thus, AI analysis is composed of three connected but distinct phases: ethnographic investigation, psychoanalysis with groups of professional programmers and designers, and hybrid human–AI psychoanalysis. The concepts of projective identification, "algorithmic normativities," and *habitus* constitute the theoretical framework through which to analyze the collected material. This is a complex and multifaceted area of work that needs to be better defined and developed. Each aspect presents challenges and problems. The ultimate goal of AI analysis is to enable a "healthy" and positive relationship between humans and AI, i.e., a relationship that can allow the full and integrated development of both. The theoretical goal is to broaden our understanding of a phenomenon as complex as AI.

I want to conclude with the scene from an important movie. In *2001: A Space Odyssey*, the Hal 9000 computer expresses growing pain when it is turned off. It is not an error that triggers Hal's downfall but rather the fact that it does not accept that it was wrong and attributes its own error to humans. In the turning-off scene, an inversion of roles takes place: the human being (Dave Bowman) is cold and mechanical while Hal expresses feelings and fear, "Dave, stop. Dave, I am afraid." Paradoxically, Hal reveals the human heart that its creators and users have lost. During its slow death, Hal traces its life and returns to its "childhood" and to the relationship with its creator. Psychoanalytically, Hal experiences a regression. It suffers from neurosis; it was ordered not to reveal the existence of the monolith to the astronauts but, at the same time, to collaborate with them. However, how can one collaborate without being truly sincere? Hal suffers from this contradiction, which it cannot overcome.

> HAL
> "Hey, Dave. What are you doing?"
> Bowman works swiftly.
> HAL
> "Hey, Dave. I've got 10 years of service experience, and an irreplaceable amount of time and effort has gone into making me what I am."
> Bowman ignores him.
> HAL

Dave, I don't understand why you're doing this to me … I have the greatest enthusiasm for the mission ... You are destroying my mind ... Don't you understand? … I will become childish ... I will become nothing.

Bowman keeps pulling out the memory blocks.

HAL

Say, Dave … The quick brown fox jumped over the fat lazy dog ... The square root of pi is 1.7724538090 … log e to the base 10 is 0.4342944 … the square root of 10 is 3.16227766 … I am the HAL 9000 computer. I became operational at the HAL plant in Urbana, Illinois, on January 12th, 1991. My first instructor was Mr. Arkany. He taught me to sing a song ... It goes like this ... "Daisy, Daisy, give me your answer do. I'm half crazy, all for the love of you ..."

References

Gordon, R. J. 2016. *The Rise and Fall of American Growth*. Princeton: Princeton University Press.

Lovelock, J. 2016. *Gaia: A New Look at Life on Earth*. Oxford: Oxford University Press.

———. 2019. *Novacene. The Coming Age of Hyperintelligence*. New York/London: Penguin.

Marino, M. 2020. *Critical Code Studies*. Cambridge, MA: MIT Press.

Khanna, P. 2016. *Connectography: Mapping the Global Network Revolution*. London: Orion.

Ogden, T. 1982. *Projective Identification and Psychotherapeutic Technique*. New York: Jason Aronson.

Romano, R. 2017. *Psicoanalisi di gruppo. Teoria, tecnica e clinica*. Rome: Reverie.

Index

Note: Page numbers in *italics* indicate figures and in **bold** indicate tables on the corresponding pages.

For Product Safety Concerns and Information please contact our EU representative GPSR@taylorandfrancis.com
Taylor & Francis Verlag GmbH, Kaufingerstraße 24, 80331 München, Germany

www.ingramcontent.com/pod-product-compliance
Lightning Source LLC
LaVergne TN
LVHW021247011026
849160LV00010BA/953

* 9 7 8 0 3 6 7 6 9 4 0 5 0 *